QUANTITATIVE
GEOCHEMISTRY

T0324795

QUANTITATIVE GEOCHEMISTRY

Haibo Zou

University of California, Los Angeles, USA
and Auburn University, USA

Imperial College Press

Published by

Imperial College Press
57 Shelton Street
Covent Garden
London WC2H 9HE

Distributed by

World Scientific Publishing Co. Pte. Ltd.
5 Toh Tuck Link, Singapore 596224
USA office: 27 Warren Street, Suite 401-402, Hackensack, NJ 07601
UK office: 57 Shelton Street, Covent Garden, London WC2H 9HE

British Library Cataloguing-in-Publication Data
A catalogue record for this book is available from the British Library.

Cover image: Parinacota Volcano, Chile. Photographed by John Hora.

ISBN-13 978-1-86094-646-2
ISBN-10 1-86094-646-1

Printed in Singapore.

In loving memory of my father,
Guilin Zou

Preface

Modern geochemistry utilizes three powerful tools: (major and trace) elements, isotopes, and equations, to study various Earth and environmental processes. A combination of the experimental tools (elements and isotopes) with theoretical tools (equations) provides penetrating insights into the Earth and environmental processes. The aim of this book is to link equations more closely with geochemical measurements, including elemental abundances and (radiogenic, radioactive and stable) isotopic compositions. The importance to use equations in scientific research has been best stated by Albert Einstein, "Equations are more important to me, because politics is for the present, but *an equation is something for eternity.*"

The book covers many exciting aspects of geochemistry where equations have been useful. Chapters 1 to 5 introduce the behaviors of trace elements and uranium-series isotope disequilibria during partial melting. Chapter 6 describes the behaviors of trace elements during crystallization, assimilation and mixing. All the above 6 chapters deal with forward modeling of geochemical processes while Chapter 7 introduces inverse geochemical modeling. Chapters 8 and 9 describe the treatments of geochemical data using error analysis and least-square fitting. These treatments are applicable not only to geochemical data but also to experimental data in other scientific fields. Chapters 10 and 11 are concerned with mass spectrometry (for the measurement of isotopic compositions using mass spectrometers) and clean lab chemistry, with emphasis on mass fractionation in ionization processes and single and double spike isotope dilution methods. Chapter 12 provides some fun in Pb isotope modeling from an uncommon perspective. For example, the well-known zircon Concordia plots are treated with the second derivatives. Chapter 13 describes geochemical kinematics and dynamics and emphasizes the combination of geochemistry with physics.

Although the mathematical methods in this book include algebra, calculus, differential equations, matrix, statistics, and numerical analyses, students with background in algebra and calculus alone are able to understand most of the contents. In addition, since simple models are presented before more complex models and additional parameters are added gradually, students should not worry about the difficulties in mathematics.

The book is problem-oriented, and thus does not cover everything in geochemistry. It is hoped that the book may help enhance students' abilities to understand, criticize, appreciate, and apply available models for deep insights of fundamental geochemical problems. For some students inspired to explore further, they may develop creative models of their own, on the basis of a full understanding of the model development in this book first.

In addition to the author's published and unpublished theoretical work, the author heavily drew upon the quantitative models of more than a hundred scientists listed in the references. I hope that their scientific contributions are correctly acknowledged in the text, and I wish to express my sincere thanks to all of them. Note that many equations in geochemistry were originally derived by geochemists (in a broad sense) themselves, not simply borrowed from chemistry, physics, or engineering literature.

It is also a great pleasure to thank many people individually for their help. First and foremost, I would like to express my deepest gratitude to Kevin D. McKeegan for his enduring inspiration, support, generosity, and mentoring. Sincere thanks also go to Mark T. Harrison for his valuable comments and corrections on the manuscript and to Jerry Wasserburg for his encouragement "keep doing science". I have greatly benefited from interactions with An Yin, John Wasson, Fred Frey, Alan Zindler, Mary Reid, Asish Basu, Vincent Salters, Roy Odom, David D. Loper, Sam Mukasa, Craig Manning, Ken Sims and late Denis M. Shaw. I learned from An Yin and John Wasson to add physics to geochemical research. I am grateful to Asish Basu for bringing me to USA for graduate studies 13 years ago. John Hora generously provided the fantastic cover image of the Parinacota volcano. I am thankful to Hongwu Xu, Huifang Xu, Ying Ma, Mariana Cosarinsky, and Yingfeng Xu for their help with pictures and references. I am indebted to Chee-Hok Lim for efficient editing, to Jimmy Low for excellent cover design, and to Imperial College Press Chairman Kok Khoo Phua for book invitation. Last, but not the least, I thank my wife Yuhong Wang and my daughter Joy Zou for their support and encouragement, without which this book would not have been completed.

I'd deeply appreciate hearing about any errors, criticisms, suggestions, difficulties encountered, and perhaps gentle word of encouragement. The author's email address is hzou@ess.ucla.edu. Additional supporting materials related to this book will be posted on the author's home page: http://www2.ess.ucla.edu/~hzou.

Haibo Zou

Contents

Chapter 1

Batch Melting

1.1. Overview of Melting Models

Partial melting of Earth and planetary materials is a fundamental process that contributes to the differentiation and evolution of the Earth. Modeling of partial melting using trace element concentrations is often required to understand the melt generation and segregation process and to interpret the chemical composition of primary melts. The behaviors of trace elements during partial melting present a good problem for the application of mathematics.

There are three general models based on the extent of chemical equilibrium between the solid and melt: batch, fractional (Schilling and Winchester, 1967; Gast, 1968; Shaw, 1970) and dynamic melting (Langmuir et al., 1977; McKenzie, 1985; Zou, 1998; Zou, 2000; Zou and Reid, 2001). Another model is called continuous melting (Williams and Gill, 1989; Albarede, 1995) or critical melting (Maaloe, 1982; Sobolev and Shimizu, 1992). Although continuous melting is commonly distinguished from dynamic melting in that in the former an excess melt is removed from a static column whereas in the latter the entire melting region migrates and new fertile material is added to the column, the real difference between them is only the aggregation time required to produce the magmas. The difference in aggregation time certainly affects the activity of a short-lived radioactive nuclide in magmas, however, it will not affect the concentration of stable trace elements (Williams and Gill, 1989). Therefore, although continuous melting and dynamic melting

1

appear different conceptually, for the purpose of mathematical treatment of stable trace element fractionation, they are mathematically identical.

Among the three general models, the batch melting model assumes that melt remains in equilibrium with the solid throughout the melting event whereas the fractional melting model assumes that (1) the melt is removed from the initial source as it is formed, (2) only the last drop of melt is in equilibrium with the residue, and (3) there is no residual melt. Dynamic melting involves the retention of a critical fraction of melt in the mantle residue. During dynamic melting, when the melt mass fraction in the residue is less than the critical value for melt separation (or the critical mass porosity of the residue, Φ), there is no melt extraction (as in batch melting); when the melt fraction in the residue is greater than Φ, any infinitesimal excess melt will be extracted from the matrix.

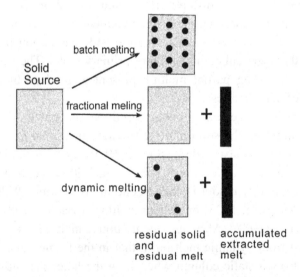

Fig. 1.1. A sample diagram showing batch, fractional and dynamic melting models. Filled circles represent melts in equilibrium with the residual solid. There is no residual melt for the fractional melting model.

The difference between the three basic models can be illustrated in the mass porosity of the melting residue (Ψ) vs. the total partial melting degree (F) diagram (Fig. 1.1). For batch melting, the mass porosity is equal to the partial melting degree until extraction of melt begins, that is, $\Psi = F$ before melt extraction begins and $\Psi = 0$ after melt extraction takes place; for (perfect) fractional melting, $\Psi = 0$ during the whole melting process; for dynamic melting, $\Psi = F$ when $F < \Phi$; $\Psi = \Phi$ when $F > \Phi$.

It is noted that the relationship between X and F is also important to distinguish the dynamic melting model from both the batch melting and the fractional melting models (Fig. 3).

For batch melting,

$$X = 0,$$

before melt extraction begins, and

$$X = F,$$

after melt extraction takes place;
for fractional melting,

$$X = F;$$

and for dynamic melting,

$$X = 0, \text{ when } F < \Phi,$$

and,

$$X = \frac{F - \Phi}{1 - \Phi}, \tag{1.1}$$

when $F > \Phi$ (Zou, 1998).

The slope in the X vs. F diagram for dynamic melting is $1/(1 - \Phi)$ and is greater than 1 because $0 < \Phi < 1$.

In this chapter, we focus on the batch melting model where melt maintains chemical equilibrium with solid, and stays with solid until the final extraction.

Define F as the degree of partial melting, that is, the mass ratio of the melt over the initial mass before melting. The balance of the total mass and the mass of an element can be presented in Table 1.1.

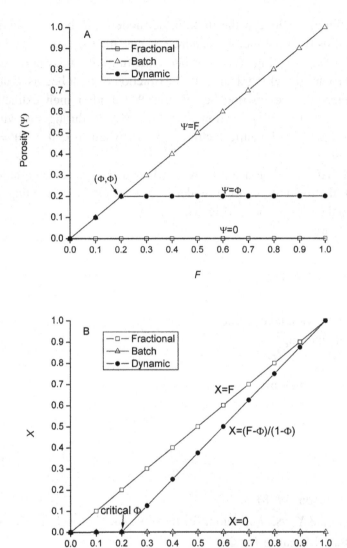

Fig. 1.2. (a) Mass porosity of melting residue (Ψ) vs. the total partial melting degree (F) diagram for batch, fractional, and dynamic melting models. (b) The fraction of extracted melt relative to the initial solid before melting starts (X) vs. the total partial melting degree (F) diagram for the three general models. Critical porosity (Φ) is set at 0.2 for illustration purpose.

Table 1.1. Mass balance during batch melting. The two-way arrows indicate interaction and equilibrium between solid and melt.

	Source	Solid	Equilibrium Melt
Conservation of total mass	M_0	$M_0(1-F)$ $\longleftrightarrow$	$M_0 F$
Conservation of the mass of an element	$C_0 M_0$	$C_s M_0(1-F)$ $\longleftrightarrow$	$C_L M_0 F$

Mass balance requirement gives

$$C_L F M_0 + C_s(1-F)M_0 = C_0 M_0. \qquad (1.2)$$

The concentration of the melt (C_L) in equilibrium with the a multiple-phase solid is related to the concentration in the solid (C_s) by

$$C_s = D C_L, \qquad (1.3)$$

where D is the bulk partition coefficient

$$D = \sum x^i K^i, \qquad (1.4)$$

where x^i is the mineral proportion in the residue and K^i is the mineral/melt partition coefficient.
Combination of Eq. (1.2) with (1.3) yields

$$C_L = \frac{C_0}{F + D(1-F)}. \qquad (1.5)$$

This is the fundamental equation for batch melting. More complex batch melting models deal with the changes of the bulk partition coefficient D. Although this formula may also be expressed as

$$C_L = \frac{C_0}{D + F(1-D)}. \qquad (1.6)$$

Equation (1.5) is convenient if we need to consider the variation of D.

Example. For a three mineral phase with 50% olivine (ol), 30% clinopyroxene (cpx) and 20% orthopyroxene (opx), if the mineral/melt partition coefficients are $K^{ol} = 0.01$, $K^{cpx} = 0.2$ and $K^{opx} = 0.08$, the bulk partition coefficients between melt and the multi-phase solid is

$$D = \sum x^i K^i = x^{ol} K^{ol} + x^{cpx} K^{cpx} + x^{opx} K^{opx}$$
$$= 0.5 \times 0.01 + 0.4 \times 0.2 + 0.1 \times 0.08 = 0.093.$$

1.2. Modal Batch Melting

During modal melting, the melting proportion is the same as the mineral source proportion and thus x^i is constant throughout the melting process $x^i = x_0^i$. When K^i is also constant ($K^i = K_0^i$), then the bulk partition coefficient D remains constant, Eq. (1.5) becomes

$$C_L = \frac{C_0}{F + D_0(1-F)} = \frac{C_0}{D_0 + F(1-D_0)}, \qquad (1.7)$$

where

$$D_0 = \sum x_0^i K_0^i. \qquad (1.8)$$

Example. Calculate C_L/C_0 for three elements with bulk partition coefficients of 0.02, 0.10 and 1.5, respectively. From Eq. (1.7) we can make Fig. 1.3 by plotting C_L/C_0 vs. F for the three D values.

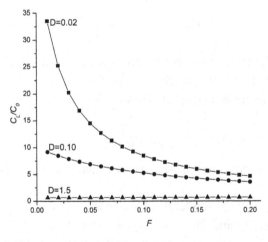

Fig. 1.3. Variation of Source-normalized concentrations with the degree of artial melting for trace elements with different bulk partition coefficients.

1.3. Nonmodal (Eutectic) Batch Melting

Modal melting normally does not happen. Due to preferential melting of some minerals in the source rock, x^i must change. For example, a mantle rock may have 10% cpx, but the cpx may contribute 25% of the melt as a result of preferential melting of cpx, resulting the decrease of x^{cpx} as a function of F during melting.

The mass conservation for the phase i gives

$$M_0\left(1-F\right)x^i + M_0 F p^i = M_0 x_0^i , \tag{1.9}$$

where p^i is the fractional contribution of phase i to the melt, x_0^i is the initial fraction of phase i in the source, $M_0\left(1-F\right)$ is the mass of the residual solid, and $M_0 F$ is the mass of the melt. Equation (1.9) may be re-expressed as

$$x^i = \frac{x_0^i - F p^i}{1-F} . \tag{1.10}$$

Note that for modal melting, melting proportion is the same as the initial mineral proportion in the source ($p^i = x_0^i$). In this case, Eq. (1.10) reduces to

$$x^i = x_0^i . \tag{1.11}$$

1.3.1. Eutectic batch melting with constant K^i

When K^i is a constant, the bulk partition coefficient becomes

$$D = \sum x^i \left(F\right) K^i = \sum \frac{x_0^i - F p^i}{1-F} K_0^i = \frac{\sum x_0^i K_0^i - F \sum p^i K_0^i}{1-F} . \tag{1.12}$$

Defining

$$P = \sum p^i K_0^i , \tag{1.13}$$

we have

$$D = \frac{D_0 - PF}{1-F} . \tag{1.14}$$

Differentiation of D with respect to F gives

$$dD/dF = (D_0 - P_0)/(1-F)^2 .$$ (1.15)

If $D_0 < P_0$, then $dD/dF < 0$ and D decreases as melting proceeds; and if $D_0 > P_0$, then $dD/dF > 0$ and D increases during melting (Fig. 1b). In fact, the relationship $D_0 < P_0$ implies the preferential melting of the minerals with high mineral/melt distribution coefficients and thus the solid residue is left with higher proportions of minerals of low distribution coefficients; the opposite is true for the case where $D_0 > P_0$. Substitution of (1.14) into (1.5) results in

$$C_L = \frac{C_0}{D_0 + F(1-P)} .$$ (1.16)

Equation (1.16) is the famous non-modal batch melting equation from Shaw (1970).

1.3.2. Eutectic batch melting with linear change of K^i

We then expand $K^i(F)$ in a Taylor series and only retain terms up to the first order

$$K^i(F) = K_0^i + a^i F .$$ (1.17)

The above simple assumption of linear variation of $K^i(F)$ as a function of F is at least appropriate for evaluating the major effects of decreasing or increasing distribution coefficients (Greenland, 1970; Hertogen and R., 1976). More complicated forms of the distribution coefficients as a function of F can also be assumed; however, they involve more parameters and do not necessarily better describe the variations in distribution coefficients.

Substitution of (1.17) and (1.10) into (1.4) gives

$$D = \sum x^i(F)K^i(F) = \sum \frac{x_0^i - Fp^i}{1-F}(K_0^i + a^i F),$$ (1.18)

or

$$D = \frac{1}{1-F}\left[-\left(\Sigma a^i p^i\right)F^2 + \left(-P_0 + \Sigma a^i x_0^i\right)F + D_0\right]. \tag{1.19}$$

Substitution of (1.19) into (1.5) gives

$$
\begin{aligned}
C_L &= \frac{C_0}{F + D(1-F)} \\
&= \frac{C_0}{D_0 + \left(1 - P_0 + \sum a^i x_0^i\right)F - \left(\sum a^i p^i\right)F^2}.
\end{aligned} \tag{1.20}
$$

1.3.3. Eutectic batch melting with linear change of K^i, linear change of p^i

If p^i is also a function of F, then Eq. (1.9) for mass conservation of phase i should be replaced by

$$x^i\left(1-F\right) = x_0^i - \int_0^F p^i dF, \tag{1.21}$$

We may assume linear variations in p^i during nonmodal melting

$$p^i = p_0^i + b^i F, \tag{1.22}$$

where $b^i = \left(p^i(F_m) - p_0^i\right)/F_m$, F_m is the maximum degree of melting when one of the phases melt completely, and $p^i(F_m)$ is the net fractional contribution to the melt when the degree of partial melting is F_m. By substituting Eq. (1.22) into Eq. (1.21), we have

$$x^i = \frac{x_0^i - F p_0^i - 0.5 b^i F^2}{1-F}. \tag{1.23}$$

Combination of Eq. (1.23), (1.17) and (1.4) leads to

$$D = \sum \frac{x_0^i - p_0^i F - 0.5 b^i F^2}{1-F}\left(K_0^i + a^i F\right), \tag{1.24}$$

or

$$D = \frac{1}{1-F}\left(A_0 F^3 + A_1 F^2 + A_2 F + D_0\right), \tag{1.25}$$

where

$$A_0 = -\left(\sum 0.5 a^i b^i\right), \tag{1.26}$$

$$A_1 = -\sum a^i p_0^i - \sum 0.5 b^i K_0^i, \tag{1.27}$$

$$A_2 = \sum a^i x_0^i - \sum p_0^i K_0^i = \sum a^i x_0^i - P_0, \tag{1.28}$$

$$D_0 = \sum K_0^i x_0^i. \tag{1.29}$$

Substitution of (1.25) into (1.5) yields

$$C_L = \frac{C_0}{D_0 + (1 + A_2)F + A_1 F^2 + A_0 F^3}. \tag{1.30}$$

1.4. Incongruent Batch Melting

Eutectic melting generates melt only, however, melting reactions in the mantle and crust often produce not only melt but also minerals. During incongruent melting, x^i changes all the time. But it varies in a different manner as compared with eutectic melting.

During partial melting of the crust and the mantle, some minerals melt congruently and others melt incongruently. Melting of congruent minerals only produces melt, and such a process can be expressed as

$$\theta_1 + \theta_2 + \cdots + \theta_u \rightarrow Liq_\theta, \tag{1.31}$$

where θ_1, θ_2, and θ_u are minerals that melt congruently and Liq_θ is the liquid formed by congruent melting.

By comparison, melting of incongruent minerals produces not only melt but also minerals, and such a melting reaction process can be expressed as

$$\alpha_1 + \alpha_2 + \cdots + \alpha_v \rightarrow \beta_1 + \beta_2 + \cdots + \beta_w + Liq_\alpha, \tag{1.32}$$

where α_1, α_2 and α_v represents the minerals that melt incongruently, β_1, β_2, and β_w are product minerals, and Liq_α is the liquid formed by melting reactions. As an example, quartz, biotite and sillimanite are α

minerals while is a β mineral, in the following melting reactions (in mass units) in a biotite-sillimanite-quartz gneiss

0.27 quartz + 0.43 biotite + 0.30 sillimanite = 0.72 cordierite + 0.28 melt.

As another example, at 10 kbar, melting of a spinel peridotite has the following melting reaction in mass units (Kinzler and Grove, 1992)

0.82 cpx +0.40 opx +0.08 spinel = 1.0 melt +0.30 olivine

clinopyroxene (cpx), orthopyroxene (opx) and spinel are α minerals whereas olivine is a β mineral.

In fact, a source rock often contains both congruent minerals and incongruent minerals, therefore, a general melting equation is

$$
\begin{aligned}
(\theta_1 + \theta_2 + \cdots + \theta_u) + (\alpha_1 + \alpha_2 + \cdots + \alpha_v) \rightarrow \\
Liq_\theta + (\beta_1 + \beta_2 + \cdots + \beta_w + Liq_\alpha)
\end{aligned}
\qquad (1.33)
$$

The minerals (β) that are produced in the melting reaction can be the congruent minerals (θ) already present in the system or minerals that are new to the system. Incongruent melting is common in the process of mantle and crust melting (Zeck, 1970; Benito-Garcia and Lopez-Ruiz, 1992; Kinzler and Grove, 1992; Kinzler, 1997; Gudfinnsson and Presnall, 1996; Walter, 1998).

Let S_θ^i and L_θ be the consumed mass of a congruent mineral and the mass of the produced melt by congruent melting, respectively. Then, according to Eq. (1.31), we have the total consumed mass of all congruent minerals as $\sum S_\theta^i = L_\theta$. Let S_α^i, S_β^i and L_α be the converted mass of an incongruent mineral, the produced mass of a mineral, and the mass of the produced melt, respectively, during incongruent melting, then, according to Eq. (1.32), we have $\sum_\alpha S_\alpha^i = \sum_\beta S_\beta^i + L_\alpha$. The mass balance for Eq. (1.33) is

$$
\sum_\alpha S_\alpha^i + \sum_\theta S_\theta^i = \sum_\beta S_\beta^i + L_\alpha + L_\theta.
\qquad (1.34)
$$

For a general melting reaction in Eq. (1.33), the degree of partial melting (F) is the mass fraction of the total melt $(L_\theta + L_\alpha)$ relative to the initial amount of the source (M_0), or,

$$F = (L_\theta + L_\alpha)/M_0 . \qquad (1.35)$$

According to the definitions of S_θ^i, S_α^i, S_β^i, L_θ, and L_α in Eq. (1.33), the total mass of the converted minerals is $\left(\sum_\theta S_\theta^i + \sum_\alpha S_\alpha^i \right)$, and the total mass of the produced melt is $(L_\theta + L_\alpha)$. Therefore, the fractional contributions of phase i to the total mass of the converted minerals through congruent melting and incongruent melting are, respectively,

$$p_\theta^i = S_\theta^i \left/ \left(\sum_\theta S_\theta^i + \sum_\alpha S_\alpha^i \right) \right. , \qquad (1.36)$$

$$p_\alpha^i = S_\alpha^i \left/ \left(\sum_\theta S_\theta^i + \sum_\alpha S_\alpha^i \right) \right. . \qquad (1.37)$$

The mass fractions of incongruent minerals converted into melt or mineral i are, respectively,

$$t^l = L_\alpha \left/ \left(\sum_\beta S_\beta^i + L_\alpha \right) = L_\alpha \left/ \sum_\alpha S_\alpha^i \right. \right. , \qquad (1.38)$$

$$t^i = S_\beta^i \left/ \left(\sum_\beta S_\beta^i + L_\alpha \right) = S_\beta^i \left/ \sum_\alpha S_\alpha^i \right. \right. . \qquad (1.39)$$

Consequently, we have

$$\sum_\theta p_\theta^i + \sum_\alpha p_\alpha^i = 1 , \qquad (1.40)$$

$$t^l + \sum_\beta t^i = 1. \qquad (1.41)$$

From (1.36), (1.37) and (1.39), we obtain the mass fraction of net converted mineral i *relative to the total converted mass* as

$$\left(S_\theta^i + S_\alpha^i - S_\beta^i\right)\bigg/\left(\sum_\theta S_\theta^i + \sum_\alpha S_\alpha^i\right) = p_\alpha^i + p_\theta^i - t^i \sum_\alpha p_\alpha^i . \quad (1.42)$$

From (1.34), (1.37), (1.39) and (1.41), we get the mass fraction of total melt *relative to the total converted mass* as

$$\left(L_\theta + L_\alpha\right)\bigg/\left(\sum_\theta S_\theta^i + \sum_\alpha S_\alpha^i\right) = 1 - \sum_\beta S_\beta^i \bigg/\left(\sum_\theta S_\theta^i + \sum_\alpha S_\alpha^i\right)$$
$$= 1 - \left(1 - t^i\right)\sum_\alpha p_\alpha^i \quad (1.43)$$

By dividing Eq. (1.42) by Eq. (1.43), we obtain the mass fraction of net converted mineral i *relative to the total produced melt*

$$q^i = \frac{S_\theta^i + S_\alpha^i - S_\beta^i}{L_\theta + L_\alpha} = \frac{p_\theta^i + p_\alpha^i - t^i \sum_\alpha p_\alpha^i}{1 - \left(1 - t^i\right)\sum_\alpha p_\alpha^i}, \quad (1.44)$$

which can be interpreted as the actual fractional contribution of mineral i to the total melt. Multiplying Eq. (1.44) by Eq. (1.35) for the definition of the degree of partial melting (F), we obtain the mass fraction of net converted mass for mineral i *relative to the initial source amount* (M_0) as

$$x_c^i = \frac{S_\theta^i + S_\alpha^i - S_\beta^i}{M_0} = Fq^i = \frac{p_\theta^i + p_\alpha^i - t^i \sum_\alpha p_\alpha^i}{1 - \left(1 - t^i\right)\sum_\alpha p_\alpha^i} F . \quad (1.45)$$

Mass balance of mineral i requires

$$x_0^i = x^i(1 - F) + x_c^i . \quad (1.46)$$

Substitution of (1.45) into (1.46) yields the variation of a mineral phase proportion as

$$x^i = \frac{x_0^i - q^i F}{1 - F}, \quad (1.47)$$

where

$$q^i = \frac{p_\theta^i + p_\alpha^i - t^i \sum_\alpha p_\alpha^i}{1 - \left(1 - t^i\right) \sum_\alpha p_\alpha^i}.$$ (1.48)

Benito-Garcia and Lopez-Ruiz (1992) derived the same equation (1.47) from a different approach. The step-by-step approach presented here is based on clear definitions and may help to better understand the parameters and terms related to the change of mineral proportions in the residual solid during incongruent melting.

Example. Assuming that this are no congruent minerals in the source, calculate q^i and x^i for the following reaction melting:

0.06 ol + 0.71 cpx + 0.23 gt = 0.52 melt +0.48 opx

(ol=olivine; cpx=clinopyroxene; gt=garnet; opx=orthopyroxene)
Based on definitions, we have

$$p_\alpha^{ol} = 0.06, \, p_\alpha^{cpx} = 0.71,$$

$$p_\alpha^{gt} = 0.06, \, t^l = 0.52,$$

$$t^{opx} = 0.48.$$

Since there are no congruent minerals we have $p_\theta^i = 0$, $\sum_\theta p_\theta^i = 0$ and $\sum_\alpha p_\alpha^i = 1$. For incongruent minerals in the reaction (ol, cpx, gt),

$$q^i = p_\alpha^i / t^l,$$

and for newly formed mineral (opx), we have

$$q^i = -t^i / t^l.$$

Thus, we have

$$q^{ol} = 0.06/0.52 = 0.12,$$

$$q^{cpx} = 0.71/0.52 = 1.37,$$

$$q^{gt} = 0.23/0.52 = 0.44,$$

$$q^{opx} = -0.48/0.52 = -0.92 .$$

The variations of the mineral proportions are

$$x^{ol} = \frac{x_0^{ol} - 0.12F}{1-F} ,$$

$$x^{cpx} = \frac{x_0^{cpx} - 1.37F}{1-F} ,$$

$$x^{gt} = \frac{x_0^{gt} - 0.44F}{1-F} ,$$

$$x^{opx} = \frac{x_0^{opx} + 0.92F}{1-F} .$$

1.4.1. Incongruent batch melting with constant K^i

For constant K^i, we have

$$K^i = K_0^i . \tag{1.49}$$

Substitution of (1.49) and (1.47) into (1.4) gives

$$D = \frac{1}{1-F}\left[D_0 - \frac{P_0 - \left(\sum_\alpha p_\alpha^i\right)\left(\sum_\beta t^i K_0^i\right)}{1-\left(1-t^l\right)\left(\sum_\alpha p_\alpha^i\right)} F \right] = \frac{D_0 - Q_0 F}{1-F} , \tag{1.50}$$

where

$$Q_0 = \sum q^i x_0^i = \frac{P_0 - \left(\sum_\alpha p_\alpha^i\right)\left(\sum_\beta t^i K_0^i\right)}{1-\left(1-t^l\right)\left(\sum_\alpha p_\alpha^i\right)} . \tag{1.51}$$

Substitution of (1.50) into (1.5) gives rise to

$$C_L = \frac{C_0}{D_0 + F(1 - Q_0)}. \qquad (1.52)$$

Example. Find Q_0 for the melting reaction

0.06 ol + 0.71 cpx + 0.23 gt = 0.52 melt +0.48 opx.

Since there are no congruent minerals in the above melting reaction, we have $\sum_\alpha p_\alpha^i = 1$. On the basis of the melting reaction, we have

$$p_\alpha^{ol} = 0.06, \, p_\alpha^{cpx} = 0.71,$$

$$p_\alpha^{gt} = 0.06, \, t^l = 0.52,$$

$$t^{opx} = 0.48.$$

Substitution of these parameters into Eq. (1.51) gives

$$
\begin{aligned}
Q_0 &= \frac{P_0 - \left(\sum_\beta t^i K_0^i\right)}{t^l} = \frac{P_0 - t^{opx} K_0^{opx}}{t^l} \\
&= \frac{p^{ol} K^{ol} + p^{cpx} K^{cpx} + p^{gt} K^{gt} - t^{opx} K^{opx}}{0.52} \\
&= \frac{0.06 K^{ol} + 0.71 K^{cpx} + 0.23 K^{gt} - 0.48 K^{opx}}{0.52}.
\end{aligned}
$$

Alternatively, Q_0 can be obtained using q^i from the example from section 1.3.1:

$$Q_0 = \sum q^i x_0^i = 0.12 K^{ol} + 1.37 K^{cpx} + 0.44 K^{gt} - 0.92 K^{opx}.$$

Example. Find Q_0 for the melting reaction

0.27 qz + 0.43 bio + 0.30 sil = 0.72 cord +0.28 melt

(qz=quartz, bio=biotite, sil=sillimanite; cord=cordirite).

On the basis of the above melting reaction, we get

$$\sum_\alpha p_\alpha^i = 1, \, p_\alpha^{qz} = 0.27,$$

$$p_\alpha^{bio} = 0.43 , \; p_\alpha^{sil} = 0.30 ,$$

$$t^l = 0.28 , \; t^{cord} = 0.72 .$$

Substitution of these parameters into Eq. (1.51) leads to

$$
\begin{aligned}
Q_0 &= \frac{P_0 - \left(\sum_\beta t^i K_0^i \right)}{t^l} = \frac{P_0 - t^{cord} K_0^{cord}}{t^l} \\
&= \frac{p^{qz} K^{qz} + p^{bio} K^{bio} + p^{sil} K^{sil} - t^{cord} K^{cord}}{0.52} \\
&= \frac{0.27 K^{ol} + 0.43 K^{cpx} + 0.30 K^{gt} - 0.72 K^{opx}}{0.28} .
\end{aligned}
$$

1.4.2. Incongruent batch melting with linear change of K^i

We may assume linear variations of distribution coefficients

$$K^i (F) = K_0^i + a^i F . \tag{1.53}$$

Combination of (1.53), (1.47) and (1.4) yields

$$
\begin{aligned}
D &= \sum x^i K^i \\
&= \sum \frac{x_0^i - F q^i}{1 - F} \left(K_0^i + a^i F \right) \tag{1.54} \\
&= \frac{1}{1 - F} \left[-\left(\Sigma a^i q^i \right) F^2 + \left(-Q_0 + \Sigma a^i x_0^i \right) F + D_0 \right] ,
\end{aligned}
$$

where

$$D_0 = \sum K_0^i x_0^i \; \text{and} \; Q_0 = \sum q^i x_0^i .$$

Substitution of Eq. (1.54) into Eq. (1.5) results in

$$C_L = \frac{C_0}{D_0 + \left(1 - Q_0 + \sum a^i x_0^i \right) F - \left(\sum a^i q^i \right) F^2} . \tag{1.55}$$

Note that if all $a^i = 0$, then Eq. (1.55) reduces to Eq. (1.52).

1.4.3. Incongruent batch melting with linear change of K^i and linear change of q^i

If q^i is also a function of F, then Eq. (1.46) for mass conservation of phase i should be replaced by

$$x^i(1-F) = x_0^i - \int_0^F q^i dF. \qquad (1.56)$$

We may assume linear variations in q^i during incongruent melting

$$q^i = q_0^i + b^i F, \qquad (1.57)$$

where $b^i = \left(p^i(F_{max}) - p_0^i\right)/F_{max}$, F_{max} is the maximum degree of melting when one of the phases melt completely, and $q^i(F_{max})$ is the net fractional contribution to the melt when the degree of partial melting is F_{max}. Substitution of (1.57) into (1.56) yields

$$x^i = \frac{x_0^i - Fq_0^i - 0.5b^i F^2}{1-F}. \qquad (1.58)$$

Combination of (1.58), (1.53) and (1.4) gives

$$D = \frac{\left[B_0 F^3 + B_1 F^2 + B_2 F + D_0\right]}{1-F}, \qquad (1.59)$$

where

$$B_0 = -\left(\sum 0.5a^i b^i\right), \qquad (1.60)$$

$$B_1 = -\sum a^i q_0^i - \sum 0.5b^i K_0^i, \qquad (1.61)$$

$$B_2 = \sum a^i x_0^i - \sum q_0^i K_0^i = \sum a^i x_0^i - Q_0, \qquad (1.62)$$

$$D_0 = \sum K_0^i x_0^i. \qquad (1.63)$$

Substitution of Eq. (1.59) into Eq. (1.5) results in

$$C_L = \frac{C_0}{D_0 + (1+B_2)F + B_1 F^2 + B_0 F^3}. \qquad (1.64)$$

Note that if all $b^i = 0$, then Eq. (1.64) reduces to Eq. (1.55). If all $a^i = 0$ and all $b^i = 0$, then Eq. (1.64) collapses to Eq. (1.52).

1.5. Summary

Table 1.2. Summary of variations of mineral proportions and bulk partition coefficients during modal, eutectic and incongruent melting.

Melting modes	Mineral proportions	Bulk partition coefficients
Modal melting	$x^i = x_0^i$	$D = \sum x_0^i K^i$
Eutectic Melting	$x^i(F) = \dfrac{x_0^i - Fp^i}{1-F}$	$D = \sum \dfrac{x_0^i - Fp^i}{1-F} K^i$
Incongruent melting	$x^i(F) = \dfrac{x_0^i - Fq^i}{1-F}$	$D = \sum \dfrac{x_0^i - Fq^i}{1-F} K^i$

Table 1.3. Relationships among different batch melting (BM) models in this chapter.

	Constant K^i	Linear K^i	Linear K^i and q^i
Incongruent Batch Melting	Incongruent BM w/ constant K^i	Incongruent BM w/ linear K^i	Incongruent BM w/ linear K^i and linear q^i
	$q^i = p^i \downarrow$	$q^i = p^i \downarrow$	$q^i = p^i \downarrow$
Eutectic Batch Melting	Eutectic BM w/ constant	Eutectic BM w/ linear K^i	Eutectic BM w/ linear K^i and linear q^i.
	$x_i = x_i^0 \downarrow$	$x_i = x_i^0 \downarrow$	$x_i = x_i^0 \downarrow$
Modal Batch Melting	Modal BM w/ constant	Modal BM w/ linear K^i	Modal BM w/ linear K^i and linear q^i

The important equations for the various batch melting models are summarized as follows:

1) Modal batch melting

$$C_L = \frac{C_0}{D_0 + F(1 - D_0)}.$$

2) Eutectic batch melting

Constant K^i and constant p^i:

$$C_L = \frac{C_0}{D_0 + F(1 - P)}.$$

Linear K^i and constant p^i:

$$C_L = \frac{C_0}{D_0 + \left(1 - P_0 + \sum a^i x_0^i\right)F - \left(\sum a^i p^i\right)F^2}.$$

Linear K^i and linear p^i:

$$C_L = \frac{C_0}{D_0 + \left(1 + A_2\right)F + A_1 F^2 + A_0 F^3},$$

where

$$A_0 = -\left(\sum 0.5 a^i b^i\right),$$

$$A_1 = -\sum a^i p_0^i - \sum 0.5 b^i K_0^i,$$

$$A_2 = \sum a^i x_0^i - \sum p_0^i K_0^i = \sum a^i x_0^i - P_0.$$

3) Incongruent batch melting

Constant K^i and constant q^i:

$$C_L = \frac{C_0}{D_0 + F(1-Q_0)}.$$

Linear K^i and constant q^i:

$$C_L = \frac{C_0}{D_0 + \left(1 - Q_0 + \sum a^i x_0^i\right)F - \left(\sum a^i q^i\right)F^2}.$$

Linear K^i and linear q^i:

$$C_L = \frac{C_0}{D_0 + \left(1 + B_2\right)F + B_1 F^2 + B_0 F^3},$$

where

$$B_0 = -\left(\sum 0.5 a^i b^i\right),$$

$$B_1 = -\sum a^i q_0^i - \sum 0.5 b^i K_0^i,$$

$$B_2 = \sum a^i x_0^i - \sum q_0^i K_0^i = \sum a^i x_0^i - Q_0.$$

References

Albarede, F., 1995. Introduction to Geochemical Modeling. Cambridge Univ. Press, Cambridge, 543 pp.

Gast, P.W., 1968. Trace element fractionations and the origin of tholeiitic and alkaline magma tpes. Geochim. Cosmochim. Acta, 32: 1057-1086.

Greenland, L.P., 1970. An equation for trace element distribution during magmatic crystallization. Am. Mineral., 55: 455-465.

Hertogen, J. and R., G., 1976. Calculation of trace element fractionation during partial melting. Geochim. Cosmochim. Acta, 40: 313-322.

Langmuir, C.H., Bender, J.F., Bence, A.E., Hanson, G.N. and Taylor, S.R., 1977. Petrogenesis of basalts from the FAMOUS-area, Mid-Atlantic ridge. Earth Planet. Sci. Lett., 36: 133-156.

McKenzie, D., 1985. ^{230}Th- ^{238}U disequilibrium and the melting processes beneath ridge axes. Earth Planet. Sci. Lett., 72: 149-157.

Schilling, J.G. and Winchester, J.W., 1967. Rare-earth fractionation and magmatic processes. In: S.K. Runcorn (Editor), Mantle of the Earth and Terrestrial Planets. Interscience Publ., pp. 267-283.

Shaw, D.M., 1970. Trace element fractionation during anatexis. Geochim. Cosmochim. Acta, 34: 237-243.

Williams, R.W. and Gill, J.B., 1989. Effect of partial melting on the uranium decay series. Geochim. Cosmochim. Acta, 53: 1607-1619.

Zou, H.B., 1998. Trace element fractionation during modal and nonmodal dynamic melting and open-system melting: A mathematical treatment. Geochim. Cosmochim. Acta, 62: 1937-1945.

Zou, H.B., 2000. Modeling of trace element fractionation during non-modal dynamic melting with linear variations in mineral/melt distribution coefficients. Geochim. Cosmochim. Acta, 64: 1095-1102.

Zou, H.B. and Reid, M.R., 2001. Quantitative modeling of trace element fractionation during incongruent dynamic melting. Geochim. Cosmochim. Acta, 65: 153-162.

Chapter 2

Fractional Melting

During batch melting, melt always remains in equilibrium with solid. In contrast, during fractional melting, melt is extracted as soon as it is generated and only the last drop of extracted melt is in equilibrium with the solid.

Table 2.1. Mass balance during fractional melting. Note that, in contrast with batch melting, there are no exchanges between the extracted melt and the residual solid (no two-way arrows in the table).

	Source	Residual Solid	Extracted Melt
Conservation of total mass	M_0	$M_0(1-F)$	$M_0 F$
Conservation of the mass of an element	$C_0 M_0$	$C_s M_0(1-F)$	$\overline{C} M_0 F$

There are several ways to derive the fundamental equation for fractional melting. Here we choose an approach that can be easily understood. The mass conservation for a trace element requires

$$C_s(1-F) + \overline{C}F = C_0 . \qquad (2.1)$$

C_s is the concentration in the solid and $\overline{C}$ is the concentration in the accumulated extracted melt and is related to the instantaneous melt C_l by

$$\overline{C} = \frac{1}{F} \int C_l dF .\qquad(2.2)$$

The solid is in equilibrium with the instantaneous melt, or, the last drop of the extracted (aggregated) melt by the following relationship

$$C_s = C_l D ,\qquad(2.3)$$

and D is the bulk partition coefficient.

Note that the residual solid is in equilibrium with instantaneous melt, rather than extracted melt. The relationship between $\overline{C}$ and C_l can be rewritten as

$$C_l = \frac{d(\overline{C}F)}{dF} .\qquad(2.4)$$

By combining Eqs. (2.2), (2.3), and (2.4), we obtain the differential equation

$$D(1-F)\frac{d(\overline{C}F)}{dF} + \overline{C}F = C_0 ,\qquad(2.5)$$

or

$$\frac{d(\overline{C}F)}{C_0 - \overline{C}F} = \frac{1}{D}\frac{dF}{1-F} .\qquad(2.6)$$

This is a fundamental equation for fractional melting. D can remain constant or change with F. This fundamental equation for fractional melting is a differential equation while that for batch melting is an algebraic equation. If D changes, it is more difficult to derive the equations for fractional melting than batch melting due to the necessity of solving differential equations for fractional melting.

2.1. Modal Fractional Melting with Constant D

Assuming D is a constant, $D = D_0$, the solution to (2.6) for the extracted fractional melt is

$$\overline{C} = \frac{C_0}{F}\left[1 - (1-F)^{1/D_0}\right].\qquad(2.7)$$

For instantaneous melt, we get

$$C_f = \frac{\partial\left(F\overline{C}\right)}{\partial F} = \frac{C_0}{D_0}\left(1 - F\right)^{(1/D_0)-1}. \tag{2.8}$$

For residual solid, we have

$$C_s = D_0 C_f = C_0\left(1 - F\right)^{(1/D_0)-1}. \tag{2.9}$$

It can be shown that Eqs. (2.7) and (2.9) satisfies the mass balance equation (2.1).

Example. Plot the variations in trace element concentrations in instantaneous melt and extracted melt for an incompatible element with $D_0 = 0.02$ during fractional melting. From Eq. (2.7) for extracted melt and Eq. (2.8) for instantaneous melt, when $D_0 = 0.02$, we can calculate the source-normalized concentrations $\overline{C}/C_0$ and C_f/C_0 at different degrees of melting in Fig. 2.1.

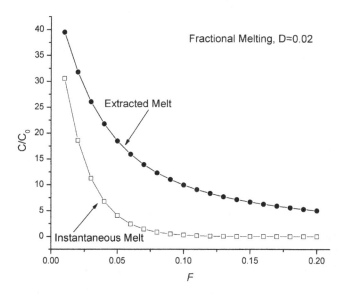

Fig. 2.1 The trace element behaviors of an incompatible element in the instantaneous fractional melt and extracted fractional melt.

Example. Compare the concentrations of a highly incompatible trace element with $D_0 = 0.02$ in the extracted melt produced by batch melting and the extracted melt generated by fractional melting.

Assuming constant bulk partition coefficient, from Eq. (1.7) for batch melting and Eq. (2.7) for fractional melting, we can make Figure 2.2 to show the behavior of a highly incompatible element with $D_0 = 0.02$ during batch melting and fractional melting.

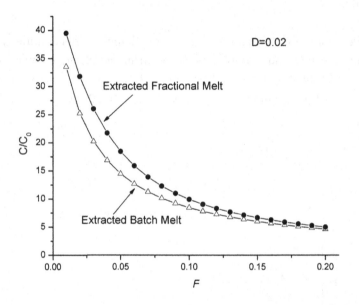

Fig. 2.2. Comparison of the concentration of an incompatible element ($D = 0.02$) in the extracted batch melt with the concentration in the extracted fractional melt in the course of partial melting.

It can be seen from Fig. 2.2 that the incompatible element concentration in the extracted melt from fractional melting is higher than that from batch melting. This is due to the fact that fractional melting is more efficient in removing incompatible elements from the source rocks. The differences between the extracted batch melt and the extracted fractional become less significant at large degrees of partial melting, for example, when $F > 20\%$.

2.2. Nonmodal (Eutectic) Fractional Melting

2.2.1. Eutectic fractional melting with constant K^i

As shown in the section 1.3 for eutectic batch melting, the variation of mineral proportions during eutectic melting is (Shaw, 1970)

$$x^i(F) = \frac{x_0^i - Fp^i}{1 - F},$$ (2.10)

and the variation of the bulk partition coefficient is

$$D = \sum \frac{x_0^i - Fp^i}{1 - F} K_0^i = \frac{\sum x_0^i K^i - \sum p^i K^i}{1 - F} = \frac{D_0 - FP}{1 - F},$$ (2.11)

where $D_0 = \sum x_0^i K_0^i$ and $P = \sum p^i K_0^i$.

Substitution of Eq. (2.11) into Eq. (2.6) yields

$$\frac{d(\overline{C}F)}{C_0 - \overline{C}F} = \frac{dF}{D_0 - PF}.$$ (2.12)

Integration of Eq. (2.12) gives

$$\overline{C} = \frac{C_0}{F}\left[1 - (1 - \frac{P}{D_0}F)^{1/P}\right].$$ (2.13)

The concentrations in the instantaneous melt is

$$C_f = \frac{\partial(\overline{C}F)}{\partial F} = \frac{C_0}{D_0}\left(1 - \frac{P}{D_0}F\right)^{\frac{1}{P}-1},$$ (2.14)

and the concentration in the solid is

$$C_s = \frac{C_0}{1-F}\left(1 - \frac{P}{D_0}F\right)^{\frac{1}{P}}.$$ (2.15)

2.2.2. Eutectic fractional melting with linear K^i

From section 1.3.2, the bulk partition coefficient for linear K^i is

$$D = \Sigma x^i\left(F\right)K^i\left(F\right) = \Sigma \frac{x_0^i - Fp^i}{1-F}\left(K_0^i + a^i F\right).$$ (2.16)

Equation (2.16) can be written as

$$D = \frac{1}{1-F}\left(AF^2 + BF + D_0\right),$$ (2.17)

where

$$A = -\left(\Sigma a^i p^i\right),$$ (2.18)

$$B = -P_0 + \Sigma a^i x_0^i.$$ (2.19)

Substitution of Eq. (2.17) into Eq. (2.6) leads to

$$\frac{d\left(F\overline{C_L}\right)}{C_0 - F\overline{C_L}} = \frac{dF}{AF^2 + BF + D_0}.$$ (2.20)

When all $a^i = 0$, then $A = 0$, and $B = -P_0$ this differential equation (2.20) reduces to (2.12). When $A \neq 0$, the solutions to Eq. (2.20) have the three cases (Hertogen and Gijbels, 1976). After corrections of their Eq. (A-1) and Eq. (A-3) as suggested by Apted and Roy (1981), the solutions are given below.

If $\Delta = B^2 - 4AD_0 > 0$, then

$$\overline{C} = \frac{C_o}{F}\left\{1 - \left|\frac{\left(2AF + B + h\right)\left(B - h\right)}{\left(2AF + B - h\right)\left(B + h\right)}\right|^{\frac{1}{h}}\right\},$$ (2.21)

$$C_f = \frac{d\left(F\overline{C}_L\right)}{dF}$$

$$= C_0 \frac{B-h}{B+h} \frac{4A}{\left(2AF+B-h\right)^2} \left|\frac{\left(2AF+B+h\right)\left(B-h\right)}{\left(2AF+B-h\right)\left(B+h\right)}\right|^{(1/h)-1},$$

$$(2.22)$$

where

$$h = \sqrt{B^2 - 4AD_0} \; ;$$

if $\Delta = 0$, then

$$\overline{C} = \frac{C_0}{F}\left[1 - \exp\left(\frac{2}{2AF+B} - \frac{2}{B}\right)\right], \qquad (2.23)$$

$$C_f = \frac{4AC_0}{\left(2AF+B\right)^2}\exp\left(\frac{2}{2AF+B} - \frac{2}{B}\right); \qquad (2.24)$$

and if $\Delta < 0$, then

$$\overline{C} = \frac{C_0}{F}\left\{1 - \exp\left[\frac{2}{k}\left(\arctan\frac{B}{k} - \arctan\frac{2AF+B}{k}\right)\right]\right\}, \quad (2.25)$$

$$C_f = \frac{4AC_0}{k^2\left[1 + \left(\frac{2AF+B}{k}\right)^2\right]}\exp\left[\frac{2}{k}\left(\arctan\frac{B}{k} - \arctan\frac{2AF+B}{k}\right)\right], \quad (2.26)$$

where

$$k = \sqrt{4AD_0 - B^2} \; .$$

2.3. Incongruent Fractional Melting

2.3.1. Incongruent fractional melting with constant K^i

From section 1.4, the variation of a mineral phase during incongruent melting is described by the following relationship

$$x^i = \frac{x_0^i - q^i F}{1 - F}, \tag{2.27}$$

where

$$q^i = \frac{p_\theta^i + p_\alpha^i - t^i \sum_\alpha p_\alpha^i}{1 - \left(1 - t^i\right)\sum_\alpha p_\alpha^i}. \tag{2.28}$$

For constant K^i, the bulk partition coefficient D is obtained by

$$D = \sum x^i \left(F\right) K^i = \sum \frac{x_0^i - F q^i}{1 - F} K_0^i = \frac{D_0 - Q_0 F}{1 - F}, \tag{2.29}$$

where

$$Q_0 = \sum q^i K_0^i = \frac{P_0 - \left(\sum_\alpha p_\alpha^i\right)\left(\sum_\beta t^i K_0^i\right)}{1 - \left(1 - t^i\right)\left(\sum_\alpha p_\alpha^i\right)}. \tag{2.30}$$

Substitution of Eq. (2.29) into (2.6) yields

$$\frac{d\left(\overline{C_L} F\right)}{C_0 - \overline{C_L} F} = \frac{dF}{D_0 - Q_0 F}. \tag{2.31}$$

The solution to Eq. (2.31) with initial condition $\overline{C_L^0} = C_0$ for the extracted melt is

$$\overline{C_L} = \frac{C_0}{F}\left\{1 - \left[1 - \frac{Q_0}{D_0} F\right]^{1/Q_0}\right\}. \tag{2.32}$$

Consequently, the concentration in the residual melt is

$$C_f = \frac{\partial \left(F\overline{C_L} \right)}{\partial F} = \frac{C_0}{D_0} \left[1 - \frac{Q_0}{D_0} F \right]^{(1/Q_0)-1}.$$ (2.33)

The concentration in the residual solid is

$$C_s = C_f D = \frac{C_0}{1-F} \left(1 - \frac{Q_0}{D_0} F \right)^{\frac{1}{Q_0}}.$$ (2.34)

2.3.2. Incongruent fractional melting with linear K^i

From section 1.4.2, the variation of the bulk partition coefficient is

$$D = \sum x^i (F) K^i (F) = \sum \frac{x_0^i - Fq^i}{1-F} \left(K_0^i + a^i F \right).$$ (2.35)

Equation (2.35) can be re-expressed as

$$D = \frac{1}{1-F} \left(AF^2 + BF + D_0 \right),$$ (2.36)

where

$$A = -\left(\Sigma a^i q^i \right),$$ (2.37)

$$B = -Q_0 + \Sigma a^i x_0^i.$$ (2.38)

Substitution of Eq. (2.36) into (2.6) yields

$$\frac{d \left(F\overline{C_L} \right)}{C_0 - X\overline{C_L}} = \frac{dF}{AF^2 + BF + D_0}.$$ (2.39)

When $A \neq 0$, the solutions to Eq. (2.39) have the following three cases. If $\Delta = B^2 - 4AD_0 > 0$, then

$$\overline{C} = \frac{C_o}{F} \left\{ 1 - \left| \frac{(2AF + B + h)(B - h)}{(2AF + B - h)(B + h)} \right|^{\frac{1}{h}} \right\},$$ (2.40)

$$C_f = \frac{B-h}{B+h} \frac{4AC_0}{(2AF+B-h)^2} \left| \frac{(2AF+B+h)(B-h)}{(2AF+B-h)(B+h)} \right|^{(1/h)-1} , \quad (2.41)$$

where $h = \sqrt{B^2 - 4AD_0}$;

if $\Delta = 0$, then

$$\overline{C} = \frac{C_0}{F}\left[1 - \exp\left(\frac{2}{2AF+B} - \frac{2}{B}\right)\right], \quad (2.42)$$

$$C_f = \frac{4AC_0}{(2AF+B)^2}\exp\left(\frac{2}{2AF+B} - \frac{2}{B}\right); \quad (2.43)$$

and if $\Delta < 0$, then

$$\overline{C} = \frac{C_0}{F}\left\{1 - \exp\left[\frac{2}{k}\left(\arctan\frac{B}{k} - \arctan\frac{2AF+B}{k}\right)\right]\right\}, \quad (2.44)$$

$$C_f = \frac{4AC_0}{k^2}\frac{1}{1+\left(\frac{2AF+B}{k}\right)^2}\exp\left[\frac{2}{k}\left(\arctan\frac{B}{k} - \arctan\frac{2AF+B}{k}\right)\right], \quad (2.45)$$

where $k = \sqrt{4AD_0 - B^2}$.

2.3.3. Incongruent fractional melting with linear K^i and linear q^i

From section 1.4.3, the variation of mineral proportion during incongruent melting with linear q^i is

$$x^i = \frac{x_0^i - q_0^i F - 0.5b^i F^2}{1-F}. \quad (2.46)$$

The bulk partition coefficient is

$$D = \sum \frac{x_0^i - q_0^i F - 0.5b^i F^2}{1-F}\left(K_0^i + a^i F\right). \quad (2.47)$$

Equation (2.47) can be written as

$$D = \frac{1}{1-F}\left(A_0 F^3 + A_1 F^2 + A_2 F + D_0\right), \qquad (2.48)$$

where

$$A_0 = -\left(\sum 0.5 a^i b^i\right), \qquad (2.49)$$

$$A_1 = -\sum a^i q_0^i - \sum 0.5 b^i K_0^i, \qquad (2.50)$$

$$A_2 = \sum a^i x_0^i - \sum q_0^i K_0^i, \qquad (2.51)$$

$$A_3 = D_0 = \sum K_0^i x_0^i. \qquad (2.52)$$

Combination of Eq. (2.48) and Eq. (2.6) yields

$$\frac{d\left(\overline{C_L}F\right)}{C_0 - \overline{C_L}F} = \frac{dF}{A_0 F^3 + A_1 F^2 + A_2 F + A_3}. \qquad (2.53)$$

Equation (2.53) can be solved numerically by the Runge-Kutta method (see Appendix 2A). Explicit solutions of (2.53) are given here.
When all $b^i = 0$, then $A_0 = 0$, and Eq. (2.53) reduces to Eq. (2.39). When $A_0 \neq 0$, there are three cases and the solutions to Eq. (2.53) depend on the discriminant of the cubic polynomial in the denominator of the right-hand side of Eq. (2.53),

$$\Delta = A_0^2 A_3^2 - 6 A_0 A_1 A_2 A_3 + 4 A_0 A_2^3 + 4 A_1^3 A_3 - 3 A_1^2 A_2^2. \qquad (2.54)$$

(1) If $\Delta > 0$, the cubic polynomial $A_0 X^3 + A_1 X^2 + A_2 X + A_3$ has one real root r_1 and two complex conjugate roots r_2 and r_3, and Eq. (2.53) can be rewritten as

$$\frac{d\left(\overline{C_L}F\right)}{C_0 - \overline{C_L}F} = u\frac{dF}{X - r_1} + v\frac{(2F + B)dF}{F^2 + BF + \eta} + w\frac{dF}{F^2 + BF + \eta}, \qquad (2.55)$$

where $F^2 + BF + \eta = (F - r_2)(F - r_3)$, and r_1, r_2 and r_3 can be obtained using the Cardan (or Tartaglia's) method. The solution to Eq. (2.55) is

$$\overline{C_L} = \frac{C_0}{F}\left(1 - \left|\frac{r_1}{F-r_1}\right|^u \left|\frac{\eta}{F^2+FX+\eta}\right|^v \exp\left[\frac{2w}{\lambda}\left(\begin{array}{c}\arctan\dfrac{B}{\lambda}\\[4pt]-\arctan\dfrac{2F+B}{\lambda}\end{array}\right)\right]\right), \quad (2.56)$$

where $\lambda = \sqrt{4\eta - B^2}$.

(2) If $\Delta = 0$, the cubic polynomial has one single real root r_1 and a double real root r_2, then

$$\frac{d\left(\overline{C_L}F\right)}{C_0 - \overline{C_L}F} = u\frac{dF}{F-r_1} + v\frac{dF}{F-r_2} + w\frac{dF}{(F-r_2)^2}, \qquad (2.57)$$

where r_1 and r_2 can be obtained using the Cardan's method. The solution to Eq. (2.57) is

$$\overline{C_L} = \frac{C_0}{F}\left(1 - \left|\frac{r_1}{F-r_1}\right|^u \left|\frac{r_2}{F-r_2}\right|^v \exp\left[\frac{wF}{r_2(F-r_2)}\right]\right). \qquad (2.58)$$

(3) If $\Delta < 0$, the cubic polynomial has three different real roots r_1, r_2 and r_3, then

$$\frac{d\left(\overline{C_L}F\right)}{C_0 - \overline{C_L}F} = u\frac{dF}{F-r_1} + v\frac{dF}{F-r_2} + w\frac{dF}{F-r_3}, \qquad (2.59)$$

where r_1, r_2 and r_3 can be obtained using the Trigonometrical method for the cubic polynomial. The solution to Eq. (2.59) is

$$\overline{C_L} = \frac{C_0}{F}\left(1 - \left|\frac{r_1}{F-r_1}\right|^u \left|\frac{r_2}{F-r_2}\right|^v \left|\frac{r_3}{F-r_3}\right|^w\right). \qquad (2.60)$$

2.4. Summary of Equations

Fundamental equation for fractional melting:

$$\frac{d(\overline{C}F)}{C_0 - \overline{C}F} = \frac{1}{D}\frac{dF}{1-F}.$$

1) Modal fractional melting with constant D

$$\overline{C} = \frac{C_0}{F}\left[1 - (1-F)^{1/D_0}\right], \qquad C_f = \frac{C_0}{D_0}(1-F)^{(1/D_0)-1}.$$

2) Eutectic fractional melting with constant K^i

$$\overline{C} = \frac{C_0}{F}\left[1 - (1 - \frac{P}{D_0}F)\right]^{1/P}, \qquad C_f = \frac{\partial(\overline{C}F)}{\partial F} = \frac{C_0}{D_0}\left(1 - \frac{P}{D_0}F\right)^{\frac{1}{P}-1}.$$

3) Incongruent fractional melting

Constant K^i

Differential equation:

$$\frac{d\left(\overline{C_L}F\right)}{C_0 - \overline{C_L}F} = \frac{dF}{D_0 - Q_0 F}.$$

Solution:

$$\overline{C} = \frac{C_0}{F}\left\{1 - \left[1 - \frac{Q_0}{D_0}F\right]^{1/Q_0}\right\}, \qquad C_f = \frac{C_0}{D_0}\left[1 - \frac{Q_0}{D_0}F\right]^{(1/Q_0)-1}.$$

Linear K^i

Differential equation:

$$\frac{d\left(F\overline{C_L}\right)}{C_0 - X\overline{C_L}} = \frac{dF}{AF^2 + BF + D_0} \, .$$

Solutions depend on $\Delta = B^2 - 4AD_0$. Analytical solutions for $\Delta > 0$, $\Delta = 0$, and $\Delta < 0$ are given in (2.40), (2.42) and (2.44), respectively.

Linear K^i and linear q^i

Differential equation:

$$\frac{d\left(\overline{C_L}F\right)}{C_0 - \overline{C_L}F} = \frac{dF}{A_0F^3 + A_1F^2 + A_2F + A_3} \, .$$

The differential equation can be solved numerically by the Runga-Kutta method. Analytical solutions depend on

$$\Delta = A_0^2 A_3^2 - 6A_0 A_1 A_2 A_3 + 4A_0 A_2^3 + 4A_1^3 A_3 - 3A_1^2 A_2^2 \, .$$

The solutions for $\Delta > 0$, $\Delta = 0$, and $\Delta < 0$ are given in (2.56), (2.58) and (2.60), respectively.

The relationships among various fractional melting models are summarized in Table 2.2.

Table 2.2. Relationships among different fractional melting (FM) models.

	Constant K^i	Linear K^i	Linear K^i and linear q^i
Incongruent Fractional	Incongruent FM w/ constant K^i	Incongruent FM w/ linear K^i	Incongruent FM w/ linear K^i and linear q^i
	$q^i = p^i \downarrow$	$q^i = p^i \downarrow$	$q^i = p^i \downarrow$
Eutectic Fractional	Eutectic FM w/ constant	Eutectic FM w/ linear K^i	Eutectic FM w/ linear K^i and linear q^i
	$x_i = x_i^0 \downarrow$	$x_i = x_i^0 \downarrow$	$x_i = x_i^0 \downarrow$
Modal Fractional	Modal FM w/ constant	Modal FM w/ linear K^i	Modal FM w/ linear K^i and linear q^i

Appendix 2A. Runge-Kutta Method

The Runga-Kutta method is an effective numerical method of solving differential equations. Consider the following first-order differential equation

$$y = F(x, y),$$

with initial condition $\qquad y(x_0) = y_0$.

According to Taylor's expansion series, we have

$$y_{i+1} = y(x_{i+1}) = y(x + \Delta x)$$
$$= y(x_i) + (\Delta x)y'(x_i) + \tfrac{1}{2!}(\Delta x)^2 y''(x_i) + \ldots$$

Since

$$y''(x_i) = \frac{d}{dx}\left(y'(x_i)\right)_{x=x_i} = \frac{d}{dx}\left(F(x, y)\right)_{x=x_i}$$
$$= \left(\frac{\partial F}{\partial x} + \frac{\partial F}{\partial y}\frac{\partial y}{\partial x}\right)_{x=x_i} = \left(\frac{\partial F}{\partial x} + \frac{\partial F}{\partial y}y'\right)_{x=x_i}$$
$$= \frac{\partial F}{\partial x}(x_i, y_i) + \frac{\partial F}{\partial y}(x_i, y_i)F(x_i, y_i),$$

we have

$$y_{i+1} = y(x_i) + (\Delta x)F(x_i, y_i) + \tfrac{1}{2!}(\Delta x)^2 \left(\frac{\partial F}{\partial x} + \frac{\partial F}{\partial y}F\right)_{(x_i, y_i)},$$

$$i = 0, 1, 2, \ldots n-1.$$

Algorithm can be rewritten in the following way:

$$y_{i+1} = y(x_i) + \tfrac{1}{2}(\Delta x)F(x_i, y_i) + \tfrac{1}{2}(\Delta x)\left(\tfrac{1}{2}F + (\Delta x)\frac{\partial F}{\partial x} + (\Delta x)\frac{\partial F}{\partial y}F\right)_{(x_i, y_i)}.$$

According to Taylor's formula, we have, for the terms in Δx

$$\frac{1}{2}F + (\Delta x)\frac{\partial F}{\partial x} + (\Delta x)\frac{\partial F}{\partial y}F = F(x + \Delta x, y_i + (\Delta x)F(x_i, y_i)).$$

Thus, we obtain the second order Runge-Kutta algorithm:

$$y_{i+1} = y(x_i) + \tfrac{1}{2}(\Delta x)\left[F(x_i, y_i) + F(x_i + \Delta x, y_i + \Delta x)F(x_i, y_i)\right],$$

or

$$y_{i+1} = y(x_i) + \tfrac{1}{2}(\Delta x)\left[M_1 + M_2\right],$$

where

$$M_1 = F(x_i, y_i),$$

$$M_2 = F(x_i + \Delta x, y_i + (\Delta x)F(x_i, y_i)).$$

The most commonly used the Runge-Kutta method is that of fourth order, consisting of the following algorithm:

$$y_{i+1} = y(x_i) + \tfrac{1}{6}(\Delta x)\left(N_1 + 2N_2 + 2N_3 + N_4\right), \qquad (2.61)$$

where

$$N_1 = F(x_i, y_i),$$

$$N_2 = F\left(x_i + \frac{1}{2}\Delta x, y_i + \frac{1}{2}(\Delta x)N_1\right),$$

$$N_3 = F\left(x_i + \frac{1}{2}\Delta x, y_i + \frac{1}{2}(\Delta x)N_2\right),$$

$$N_4 = F\left(x_i + \Delta x, y_i + (\Delta x)N_3\right).$$

An Application using the Runge-Kutta method

Find $y(0.2)$ from the first order differential equation $\dfrac{dy}{dx} = -\dfrac{x}{y}$, with the initial condition: $y(0) = 1$.

From initial condition, we have

$$x_0 = 0, \quad y_0 = 1.$$

If we use step $\Delta x = 0.2$, then we get

$$N_1 = F(x_0, y_0) = -\frac{0}{1} = 0,$$

$$N_2 = F\left(x_i + \frac{1}{2}\Delta x, y_i + \frac{1}{2}(\Delta x)N_1\right)$$

$$= -\frac{0 + 0.5 \times 0.2}{1 + 0.5 \times 0.2 \times 0}$$

$$= -0.1,$$

$$N_3 = F\left(x_i + \frac{1}{2}\Delta x, y_i + \frac{1}{2}(\Delta x)N_2\right)$$

$$= -\frac{0 + 0.5 \times 0.2}{1 + 0.5 \times 0.2 \times (-0.1)}$$

$$= -0.10101,$$

$$N_4 = F\left(x_i + \Delta x, y_i + (\Delta x)N_3\right)$$

$$= -\frac{0 + 0.2}{1 + 0.2 \times (-0.10101)}$$

$$= -0.20412.$$

Using the initial condition $y(x_0) = 1$ and the step $\Delta x = 0.2$, and substituting N_1, N_2, N_3 and N_4 into Eq. (2.61), we get

$$y_1 = y(x_0) + \tfrac{1}{6}(\Delta x)\left(N_1 + 2N_2 + 2N_3 + N_4\right)$$

$$= 1 + \frac{0.2}{6}\left(0 + 2*(-0.1) + 2*(-0.10101) - 0.20412\right)$$

$$= 0.979795.$$

We can compare the above numerical result from the Runga-Kutta method with the exact analytical solution. The exact solution to $dy/dx = -x/y$, $y(0) = 1$ is

$$y = \sqrt{1 - x^2},$$

and yields $y(0.2) = 0.979796$.

The numerical result from the Runge-Kutta method (0.979795) using step $\Delta x = 0.2$ is different from the exact solution (0.979796) only in the sixth digit.

We may certainly use a smaller step, for example, $\Delta x = 0.1$, to calculate $y(0.2)$. For the step $\Delta x = 0.1$, we can calculate $y(0.1)$ for the first step using the Runge-Kutta method (Eq. (2.61)) and the initial condition $y(0) = 1$. For the second step, we can calculate $y(0.2)$ from $y(0.1)$ and Eq. (2.61). However, for this application, since the step $\Delta x = 0.2$ is able to generate accurate value for $y(0.2)$, smaller steps are not highly necessary.

References

Apted, M.J. and Roy, S.D., 1981. Corrections to the trace element fractionation equations of Hertogen and Gijbels (1976). Geochim. Cosmochim. Acta, 45: 777-778.

Hertogen, J. and Gijbels, R., 1976. Calculation of trace element fractionation during partial melting. Geochim. Cosmochim. Acta, 40: 313-322.

Shaw, D.M., 1970. Trace element fractionation during anatexis. Geochim. Cosmochim. Acta, 34: 237-243.

Chapter 3

Dynamic Melting

The batch melting assumes no melt extraction during melting until the end whereas the fractional melting assumes any infinitesimal melt is extracted from the solid as soon as it is produced. Both models are end-member models. The reality might be in between. A more realistic scenario is that when the degree of partial melting is smaller than a critical value, the melt is in equilibrium with the solid and there is no melt extraction; when the degree of partial melting is greater than this critical value, any extra melt is extracted.

We use the physical concept of the dynamic melting model proposed by McKenzie (1985) for the situation where the rate of melting and volume porosity are constant and finite while the system of matrix and interstitial fluid is moving. This requires that the melt in excess of porosity be extracted from the matrix at the same rate at which it is formed (the details of the model are shown in Fig. 3 of McKenzie, 1985).

The essential differences between the batch melting, fractional melting and dynamic melting models have been given by (Zou, 1998) and can be summarized in Table 3.1.

Table 3.1 Comparison of three melting models.

	Batch	Fractional	Dynamic
Porosity	$= F$ before extraction $= 0$ after extraction	$= 0$	$= F$ when $F \leq \Phi$ $= \Phi$ when $F > \Phi$
X	$= 0$ before extraction $= F$ after extraction	$= F$	$= 0$ when $F \leq \Phi$ $= \dfrac{F - \Phi}{1 - \Phi}$ when $F > \Phi$

The mass balance of total mass and the mass of an element can be summarized in Table 3.2.

Table 3.2. Mass balance for dynamic melting.

	Source	Extracted Melt	Residual melt	Residual Solid
Conservation of total mass	M_0	$M_0 X$	$M_0(1-X)\Phi$	$M_0(1-X)(1-\Phi)$
Conservation of the mass of an element	$C_0 M_0$	$\overline{C} M_0 X$	$C_f M_0(1-X)\Phi$	$C_s M_0(1-X)(1-\Phi)$

Mass conservation for a trace element in dynamic melting requires

$$C_s M_0(1-X)(1-\Phi)+C_f M_0(1-X)\Phi+\overline{C}M_0 X = C_0 M_0 . \qquad (3.1)$$

The concentration in residual solid is related to the concentration in residual melt by

$$C_s = C_f D , \qquad (3.2)$$

where D is the bulk partition coefficient between solid and the residual melt. The concentration in residual melt is related to the concentration in the extracted melt by

$$C_f = \frac{d(\overline{C}X)}{dX} . \qquad (3.3)$$

Combination of (3.1), (3.2) and (3.3) yields

$$[\Phi+(1-\Phi)D](1-X)\frac{d(\overline{C}X)}{dX}+\overline{C}X = C_0 , \qquad (3.4)$$

or

$$\frac{d(\overline{C}X)}{C_0 -\overline{C}X} = \frac{1}{\Phi+(1-\Phi)D}\frac{dX}{1-X} . \qquad (3.5)$$

The term $\Phi + (1-\Phi)D$ can be regarded as effective partition coefficient between the latest drop of extracted melt and total residue (residual solid + residual melt). Eq. (3.5) is the fundamental equation for dynamic melting.

3.1. Modal Dynamic Melting with Constant D

If D is a constant, then by integrating both sides from 0 to X, we have

$$\overline{C} = \frac{C_0}{X}\left[1 - (1-X)^{1/[\Phi+(1-\Phi)D_0]}\right].$$ (3.6)

For residual melt,

$$C_f = \frac{\partial(\overline{C}X)}{\partial X} = \frac{C_0}{\Phi+(1-\Phi)D_0}(1-X)^{1/[\Phi+(1-\Phi)D_0]-1}.$$ (3.7)

For residual solid

$$C_s = D_0 C_f.$$ (3.8)

For total residue,

$$C_{res} = \Phi C_f + (1-\Phi)C_s = C_0(1-X)^{1/[\Phi+(1-\Phi)D_0]-1}.$$ (3.9)

Example. Calculate the concentration of an incompatible element normalized by the initial source concentration (C/C_0) with bulk partition coefficient $D = 0.02$ in the extracted dynamic melt and residual melt. Assuming $\phi = 0.01$.

By substituting $D = 0.02$ and $\phi = 0.01$ into Eq. (3.6), we can obtain the source-normalized concentration in the extracted melt $\overline{C}/C_0$ as a function of X. The results are plotted in Fig. 3.1. Similarly, substituting $D = 0.02$ and $\phi = 0.01$ into Eq. (3.7), we can obtain the source normalized concentration in the residual melt C_f/C_0 as a function of X. The results are also plotted in Fig. 3.1. Note that after melt extraction begins, residual melt is more depleted in incompatible elements than the extracted melt. At about 15% of melting, the residual melt does not contain much incompatible elements.

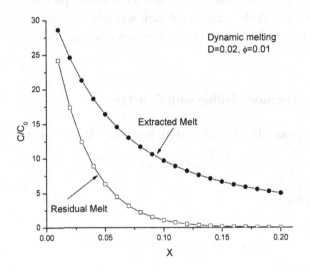

Fig. 3.1. Behaviors of an incompatible element in residual melt and extracted melt.

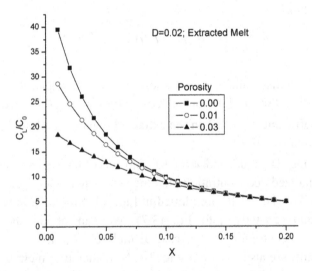

Fig. 3.2. Variation of an incompatible element concentration with $D = 0.02$ in the extracted melt during melting at different porosities. When porosity is zero, it is perfect fractional melting.

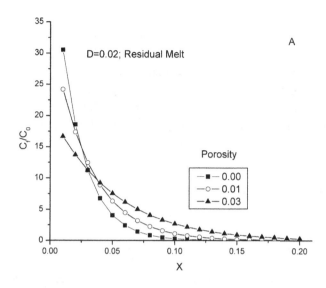

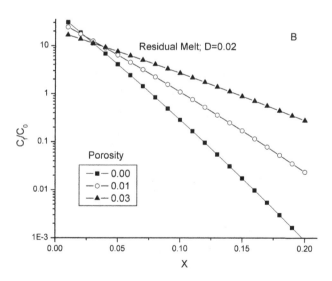

Fig. 3.3. Variations of the concentration in the residual melt as a function of the fraction of the extracted melt. (A) is for linear vertical scale (B) is for log vertical scale (to show the differences at high X).

The effects of varying critical porosity on the concentrations of incompatible elements in the extracted melt are displayed in Fig. 3.2. The incompatible element concentrations in the extracted melt increase with the decrease in the critical porosity, owing to more efficient removal of incompatible elements at lower critical porosity.

Figure 3.3 illustrates the effects of varying critical porosity on the incompatible element concentrations in the instantaneous melt. The incompatible elements decrease more quickly with the decrease of the critical porosity.

3.2. Eutectic Dynamic Melting

3.2.1. Eutectic dynamic melting with constant K^i

According to Chapter 1, the variation of the mineral proportions (x^i) in the source rocks as a function of the degree of partial melting (F) is given by

$$x^i = \frac{x_0^i - Fp^i}{1-F},$$ (3.10)

where x_0^i is the initial mineral proportion and p^i is the melting proportion. The bulk partition coefficient is given by

$$D = \sum x^i K^i.$$ (3.11)

Assuming constant K^i and substituting Eq. (3.11) into Eq. (3.10), we have

$$D = \frac{D_0 - P_0 F}{1-F},$$ (3.12)

where

$$D_0 = \sum x^i K_0^i,$$ (3.13)

$$P_0 = \sum p^i K_0^i.$$ (3.14)

F is related to X by

$$F = \Phi(1-X) + X.$$ (3.15)

We can express D as a function of X

$$D = \frac{D_0 - P_0\left[X + \Phi(1-X)\right]}{(1-X)(1-\Phi)}.$$ (3.16)

Substitution of Eq. (3.16) into the fundamental Eq. (3.5) gives

$$\frac{d\left(X\overline{C_L}\right)}{C_0 - X\overline{C_L}} = \frac{dX}{\left[D_0 + \Phi(1-P_0)\right] - \left[\Phi + P_0(1-\Phi)\right]X}.$$ (3.17)

The solution to Eq. (3.17) is

$$\overline{C_L} = \frac{C_0}{X}\left\{1 - \left[1 - \frac{X\left[P_0 + \Phi(1-P_0)\right]}{D_0 + \Phi(1-P_0)}\right]^{1/\left[\Phi + (1-\Phi)P_0\right]}\right\}.$$ (3.18)

For residual melt,

$$C_f = \frac{\partial\left(X\overline{C_L}\right)}{\partial X}$$

$$= \frac{C_0}{D_0 + \Phi(1-P_0)}\left[1 - \frac{P_0 + \Phi(1-P_0)}{D_0 + \Phi(1-P_0)}X\right]^{\left\{1/\left[\Phi + (1-\Phi)P_0\right]\right\}-1}.$$ (3.19)

For residual solid,

$$C_s = DC_f.$$ (3.20)

For total residue,

$$C_{res} = \Phi C_f + (1-\Phi)C_s.$$ (3.21)

3.2.2. Eutectic dynamic melting with linear change of K^i

We assume linear variation of mineral melt partition coefficient

$$K^i = K_0^i + a^i F.$$ (3.22)

Combination of (3.22), (3.10) and (3.11) results in

$$D = \sum x^i(F)K^i(F) = \sum \frac{x_0^i - Fp^i}{1-F}\left(K_0^i + a^i F\right).$$ (3.23)

The following relationship between F and X, which is the fraction of the extracted melt relative to the initial solid, is necessary to express $D_{s/f}$ as a function of X (Zou, 1998)

$$F = X + (1 - X)\Phi.$$ (3.24)

Although the above equation can also be written as

$$F = \Phi + (1 - \Phi)X.$$ (3.25)

We prefer the expression in Eq. (3.24) because both F and X are relative to the original mass before melting but Φ is relative to the total residue. By substituting Eq. (3.24) into Eq. (3.24), we obtain

$$D = \sum \frac{x_0^i - [X + (1 - X)\Phi] p^i}{1 - [X + (1 - X)\Phi]} \{K_0^i + a^i [X + (1 - X)\Phi]\}.$$ (3.26)

If $\Phi = 0$ and $a^i = 0$ for all i, which is the case for perfect fractional melting with constant mineral/melt distribution coefficients, Eq. (3.26) reduces to the following equation of Shaw (1970):

$$D_{s/f} = (D_0 - P_0 F)/(1 - F),$$ (3.27)

where $D_0 = \sum x_0^i K_0^i$ is the initial bulk distribution coefficient and $P_0 = \sum p^i K_0^i$ is the weighted distribution coefficient of the total melt. Substitution of (3.26) into the fundamental equation (3.5) for dynamic melting gives

$$\frac{d(X\overline{C})}{C_0 - X\overline{C}} = \frac{dX}{AX^2 + BX + Q},$$ (3.28)

where

$$A = -(\Sigma a^i p^i)(1 - \Phi)^2,$$ (3.29)

$$B = -\Phi - (P_0 - \Sigma a^i x_0^i)(1 - \Phi) - 2\Phi(1 - \Phi)(\Sigma a^i p^i),$$ (3.30)

and

$$E = \Phi + D_0 - \Phi(P_0 - \Sigma a^i x_0^i) - (\Sigma a^i p^i)\Phi^2.$$ (3.31)

Note that D is related to A, B and E by

$$D = \left(\frac{AX^2 + BX + E}{1 - X} - \Phi \right) \Big/ \left(1 - \Phi \right). \tag{3.32}$$

When, all $a^i = 0$, that mineral/melt partition coefficient remain constant, then $A = 0$, $B = -\Phi - P_0 \left(1 - \Phi \right)$ and $E = D_0 + \Phi \left(1 - P_0 \right)$. Therefore, Eq. (3.28) reduces to (3.17).

When $A \neq 0$, the solutions to Eqn. 15 have the following three cases.

If $\Delta = B^2 - 4AE > 0$, then

$$\overline{C} = \frac{C_0}{X} \left\{ 1 - \left| \frac{(2AX + B + h)(B - h)}{(2AX + B - h)(B + h)} \right|^{\frac{1}{h}} \right\}, \tag{3.33}$$

$$
\begin{aligned}
C_f &= \frac{d\left(X\overline{C} \right)}{dX} \\
&= C_0 \frac{B - h}{B + h} \frac{4A}{(2AX + B - h)^2} \left| \frac{(2AX + B + h)(B - h)}{(2AX + B - h)(B + h)} \right|^{(1/h)-1},
\end{aligned}
$$

$$\tag{3.34}$$

where $h = \sqrt{B^2 - 4AE}$;

if $\Delta = 0$, then

$$\overline{C} = \frac{C_0}{X} \left[1 - \exp\left(\frac{2}{2AX + B} - \frac{2}{B} \right) \right], \tag{3.35}$$

$$C_f = \frac{4AC_0}{(2AX + B)^2} \exp\left(\frac{2}{2AX + B} - \frac{2}{B} \right); \tag{3.36}$$

and if $\Delta < 0$, then

$$\overline{C} = \frac{C_0}{X} \left\{ 1 - \exp\left[\frac{2}{k} \left(\arctan \frac{B}{k} - \arctan \frac{2AX + B}{k} \right) \right] \right\}, \tag{3.37}$$

$$C_f = \frac{4AC_0}{k^2} \frac{1}{1 + \left(\dfrac{2AX + B}{k}\right)^2} \times$$

$$\exp\left[\frac{2}{k}\left(\arctan\frac{B}{k} - \arctan\frac{2AX + B}{k}\right)\right], \qquad (3.38)$$

where $k = \sqrt{4AE - B^2}$.

For all the above three cases, the concentration in the residual solid is

$$C_s = DC_f . \qquad (3.39)$$

3.3. Incongruent Dynamic Melting

3.3.1. Incongruent dynamic melting with constant K^i

The variation of mineral proportion during incongruent melting is

$$x^i = \frac{x_0^i - q^i F}{1 - F}, \qquad (3.40)$$

where

$$q^i = \frac{p_\theta^i + p_\alpha^i - t^i \sum_\alpha p_\alpha^i}{1 - (1 - t^i) \sum_\alpha p_\alpha^i}. \qquad (3.41)$$

The bulk partition coefficient is

$$D = \sum x^i K^i . \qquad (3.42)$$

For constant $K^i = K_0^i$, combination of (3.40), (3.41) and (3.42) gives the bulk partition coefficient

$$D = \sum x^i K^i = \frac{D_0 - Q_0 F}{1 - F}, \qquad (3.43)$$

where

$$Q_0 = \frac{P_0 - \left(\sum_\alpha p_\alpha^i\right)\left(\sum_\beta t^i K_0^i\right)}{1 - \left(1 - t^I\right)\left(\sum_\alpha p_\alpha^i\right)}. \tag{3.44}$$

Substitution of Eq. (3.43) into Eq. (3.5) yields the differential equation for incongruent dynamic melting (IDM) (Zou and Reid, 2001)

$$\frac{d\left(\overline{C_L}X\right)}{C_0 - \overline{C_L}X} = \frac{dX}{\left[D_0 + \Phi\left(1 - Q_0\right)\right] - \left[Q_0 + \Phi\left(1 - Q_0\right)\right]X}. \tag{3.45}$$

The solution to Eq. (3.45) for the extracted melt is

$$\overline{C_L} = \frac{C_0}{X}\left\{1 - \left[1 - \frac{Q_0 + \Phi(1 - Q_0)}{D_0 + \Phi(1 - Q_0)}X\right]^{1/[\Phi + (1-\Phi)Q_0]}\right\}. \tag{3.46}$$

Consequently, the concentration in the residual melt is

$$C_f = \frac{\partial\left(X\overline{C_L}\right)}{\partial X}$$

$$= \frac{C_0}{D_0 + \Phi(1 - Q_0)}\left[1 - \frac{Q_0 + \Phi(1 - Q_0)}{D_0 + \Phi(1 - Q_0)}X\right]^{\{1/[\Phi + (1-\Phi)Q_0]\}-1}. \tag{3.47}$$

The concentration in the residual solid is

$$C_s = C_f D. \tag{3.48}$$

where D is given by Eq. (3.42).
The concentration in the total residue is

$$C_{res} = \Phi C_f + \left(1 - \Phi\right)C_s.$$

3.3.2. Incongruent dynamic melting with linear change of K^i

We may assume linear variations of distribution coefficients

$$K^i = K_0^i + aF. \tag{3.49}$$

Combination of (3.40), (3.41), (3.42) and (3.49) gives the bulk partition coefficient D for linear change of K^i. And the term $\Phi + (1-\Phi)D$, which is effective partition coefficient, is given by

$$\Phi + (1-\Phi)D = \frac{1}{1-X}\left(UX^2 + VX + W\right), \tag{3.50}$$

where

$$U = -\left(\Sigma a^i q^i\right)(1-\Phi)^2, \tag{3.51}$$

$$V = -\Phi - \left(Q_0 - \Sigma a^i x_0^i\right)(1-\Phi) - 2\Phi(1-\Phi)\left(\Sigma a^i q^i\right), \tag{3.52}$$

and

$$W = \Phi + D_0 - \Phi\left(Q_0 - \Sigma a^i x_0^i\right) - \left(\Sigma a^i q^i\right)\Phi^2. \tag{3.53}$$

q^i and Q_0 are given by Eq. (3.41) and (3.44), respectively.

Note that D is related to U, V and W by

$$D = \left(\frac{UX^2 + VX + W}{1-X} - \Phi\right)\Big/(1-\Phi). \tag{3.54}$$

Combination of Eq. (3.50) with the governing equation (3.5) results in

$$\frac{d\left(X\overline{C_L}\right)}{C_0 - X\overline{C_L}} = \frac{dX}{UX^2 + VX + W}. \tag{3.55}$$

When, all $a^i = 0$, in other words, $U = 0$, then from (3.51), (3.52) and (3.53), we obtain $V = -\left[Q_0 + \Phi(1-Q_0)\right]$ and $W = D_0 + \Phi(1-Q_0)$ and Eq. (3.55) reduces to Eq. (3.17).

When $U \neq 0$, the solutions to Eq. (3.55) have the following three scenarios.

If $\Delta = V^2 - 4UW > 0$, then

$$\overline{C_L} = \frac{C_o}{X}\left\{1 - \left|\frac{(2UX + V + h)(V - h)}{(2UX + V - h)(V + h)}\right|^{\frac{1}{h}}\right\}, \tag{3.56}$$

$$C_f = \frac{d\left(X\overline{C_L}\right)}{dX}$$

$$= C_0 \frac{V-h}{V+h} \frac{4U}{(2UX+V-h)^2} \left|\frac{(2UX+V+h)(V-h)}{(2UX+V-h)(V+h)}\right|^{(1/h)-1},$$

(3.57)

where $h = \sqrt{V^2 - 4UW}$;
if $\Delta = 0$, then

$$\overline{C_L} = \frac{C_0}{X}\left[1 - \exp\left(\frac{2}{2UX+V} - \frac{2}{V}\right)\right],$$

(3.58)

$$C_f = \frac{4UC_0}{(2UX+V)^2}\exp\left(\frac{2}{2UX+V} - \frac{2}{V}\right);$$

(3.59)

and if $\Delta < 0$, then

$$\overline{C_L} = \frac{C_0}{X}\left\{1 - \exp\left[\frac{2}{k}\left(\arctan\frac{V}{k} - \arctan\frac{2UX+V}{k}\right)\right]\right\},$$

(3.60)

$$C_f = \frac{4UC_0}{k^2}\frac{1}{1+\left(\frac{2UX+V}{k}\right)^2} \times$$

(3.61)

$$\exp\left[\frac{2}{k}\left(\arctan\frac{V}{k} - \arctan\frac{2UX+V}{k}\right)\right],$$

where $k = \sqrt{4UW - V^2}$.
The concentration in the solid residue is

$$C_s = DC_f,$$

(3.62)

where D is given by Eq. (3.54).

3.3.3. Incongruent dynamic melting with linear K^i and linear q^i

From section 1.4.2, the variation of bulk partition coefficient with linear K^i and linear q^i is given by

$$x^i = \frac{x_0^i - Fq_0^i - 0.5b^i F^2}{1 - F}. \tag{3.63}$$

A combination of (3.24), (3.42), (3.49) and (3.63) gives

$$D = \sum \frac{x_0^i - q_0^i \left[X + (1-X)\Phi \right] - 0.5b^i \left[X + (1-X)\Phi \right]^2}{1 - \left[X + (1-X)\Phi \right]} \left\{ \begin{array}{l} K_0^i + \\ a^i \left[X + (1-X)\Phi \right] \end{array} \right\}. \tag{3.64}$$

And the effective partition coefficient is

$$\Phi + (1-\Phi)D = \frac{1}{1-X} \left(A_0 X^3 + A_1 X^2 + A_2 X + A_3 \right), \tag{3.65}$$

where

$$A_0 = -\left(\sum 0.5a^i b^i \right)(1-\Phi)^3, \tag{3.66}$$

$$A_1 = -\sum \left[a^i \left(q_0^i + b^i \Phi \right) \right](1-\Phi)^2 \\ -\sum \left[0.5b^i \left(K_0^i + a^i \Phi \right) \right](1-\Phi)^2, \tag{3.67}$$

$$A_2 = -\Phi + \sum \left[a^i \left(x_0^i - q_0^i \Phi - 0.5b^i \Phi \right) \right](1-\Phi) \\ -\sum \left[\left(q_0^i + b^i \Phi \right)\left(K_0^i + a^i \Phi \right) \right](1-\Phi), \tag{3.68}$$

$$A_3 = \Phi + \sum \left(K_0^i + a^i \Phi \right)\left(x_0^i - q_0^i \Phi - 0.5b^i \Phi^2 \right). \tag{3.69}$$

Note that D is related to A_0, A_1, A_2 and A_3 by

$$D = \left(\frac{A_0 X^3 + A_1 X^2 + A_2 X + A_3}{1 - X} - \Phi \right) \Big/ (1-\Phi). \tag{3.70}$$

Combination of (3.65) with the governing differential equation (3.5) yields

$$\frac{d\left(\overline{C_L}X\right)}{C_0 - \overline{C_L}X} = \frac{dX}{A_0X^3 + A_1X^2 + A_2X + A_3}. \tag{3.71}$$

Equation (3.71) can be solved numerically using the Runge-Kutta method. Analytical solutions are also presented here. When all $b^i = 0$, Eq. (3.71) collapses to Eq. (3.55). When $A_0 \neq 0$, there are three cases and the solutions to Eq. (3.71)depend on the discriminant of the cubic polynomial in the denominator of the right-hand side of Eq. (3.71),

$$\Delta = A_0{}^2A_3{}^2 - 6A_0A_1A_2A_3 + 4A_0A_2{}^3 + 4A_1{}^3A_3 - 3A_1{}^2A_2{}^2. \tag{3.72}$$

(1) If $\Delta > 0$, the cubic polynomial $A_0X^3 + A_1X^2 + A_2X + A_3$ has one real root r_1 and two complex conjugate roots r_2 and r_3, and Eq. (3.71) can be written as

$$\frac{d\left(\overline{C_L}X\right)}{C_0 - \overline{C_L}X} = u\frac{dX}{X - r_1} + v\frac{(2X + B)dX}{X^2 + BX + \eta} + w\frac{dX}{X^2 + BX + \eta}, \tag{3.73}$$

where $X^2 + BX + \eta = (X - r_2)(X - r_3)$, and r_1, r_2 and r_3 can be obtained using the Cardan (or Tartaglia's) method. The solution to Eq. (3.73) is

$$\overline{C_L} = \frac{C_0}{X}\left(\begin{array}{c} 1 - \left|\dfrac{r_1}{X - r_1}\right|^u \left|\dfrac{\eta}{X^2 + BX + \eta}\right|^v \times \\ \exp\left[\dfrac{2w}{\lambda}\left(\arctan\dfrac{B}{\lambda} - \arctan\dfrac{2X + B}{\lambda}\right)\right] \end{array}\right), \tag{3.74}$$

where $\lambda = \sqrt{4\eta - B^2}$;

(2) If $\Delta = 0$, the cubic polynomial has one single real root r_1 and a double real root r_2, then

$$\frac{d\left(\overline{C_L}X\right)}{C_0 - \overline{C_L}X} = u\frac{dX}{X - r_1} + v\frac{dX}{X - r_2} + w\frac{dX}{(X - r_2)^2}, \tag{3.75}$$

where r_1 and r_2 can be obtained using the Cardan's method. The solution to Eq. (3.75) is

$$\overline{C_L} = \frac{C_0}{X}\left(1 - \left|\frac{r_1}{X - r_1}\right|^u \left|\frac{r_2}{X - r_2}\right|^v \exp\left[\frac{wX}{r_2(X - r_2)}\right]\right); \qquad (3.76)$$

(3) If $\Delta < 0$, the cubic polynomial has three different real roots r_1, r_2 and r_3, then

$$\frac{d\left(\overline{C_L}X\right)}{C_0 - \overline{C_L}X} = u\frac{dX}{X - r_1} + v\frac{dX}{X - r_2} + w\frac{dX}{X - r_3}, \qquad (3.77)$$

where r_1, r_2 and r_3 can be obtained using the Trigonometrical method for the cubic polynomial. The solution to Eq. (3.77) is

$$\overline{C_L} = \frac{C_0}{X}\left(1 - \left|\frac{r_1}{X - r_1}\right|^u \left|\frac{r_2}{X - r_2}\right|^v \left|\frac{r_3}{X - r_3}\right|^w\right). \qquad (3.78)$$

3.4. Summary of Equations

Fundamental equation for dynamic melting:

$$\frac{d(\overline{C}X)}{C_0 - \overline{C}X} = \frac{1}{\Phi + (1 - \Phi)D}\frac{dX}{1 - X}.$$

1) Modal dynamic melting with constant D

$$\overline{C} = \frac{C_0}{X}\left[1 - (1 - X)^{1/\left[\Phi + (1 - \Phi)D_0\right]}\right],$$

$$C_f = \frac{C_0}{\Phi + (1 - \Phi)D_0}(1 - X)^{1/\left[\Phi + (1 - \Phi)D_0\right] - 1}.$$

2) Eutectic dynamic melting with constant K^i

Differential Equation:

$$\frac{d\left(X\overline{C_L}\right)}{C_0 - X\overline{C_L}} = \frac{dX}{\left[D_0 + \Phi\left(1 - P_0\right)\right] - \left[\Phi + P_0\left(1 - \Phi\right)\right]X} .$$

Solution:

$$\overline{C_L} = \frac{C_0}{X}\left\{1 - \left[1 - \frac{X\left[P_0 + \Phi\left(1 - P_0\right)\right]}{D_0 + \Phi\left(1 - P_0\right)}\right]^{1/\left[\Phi + (1-\Phi)P_0\right]}\right\},$$

$$C_f = \frac{C_0}{D_0 + \Phi(1 - P_0)}\left[1 - \frac{P_0 + \Phi(1 - P_0)}{D_0 + \Phi(1 - P_0)}X\right]^{\{1/[\Phi + (1-\Phi)P_0]\}-1} .$$

3) Incongruent dynamic melting

Constant K^i

Differential equation:

$$\frac{d\left(\overline{C_L}X\right)}{C_0 - \overline{C_L}X} = \frac{dX}{\left[D_0 + \Phi\left(1 - Q_0\right)\right] - \left[Q_0 + \Phi\left(1 - Q_0\right)\right]X} .$$

Solution:

$$\overline{C_L} = \frac{C_0}{X}\left\{1 - \left[1 - \frac{Q_0 + \Phi(1 - Q_0)}{D_0 + \Phi(1 - Q_0)}X\right]^{1/\left[\Phi + (1-\Phi)Q_0\right]}\right\},$$

$$C_f = \frac{C_0}{D_0 + \Phi(1 - Q_0)}\left[1 - \frac{Q_0 + \Phi(1 - Q_0)}{D_0 + \Phi(1 - Q_0)}X\right]^{\{1/[\Phi + (1-\Phi)Q_0]\}-1} .$$

Linear K^i

Differential equation:

$$\frac{d\left(X\overline{C_L}\right)}{C_0 - X\overline{C_L}} = \frac{dX}{UX^2 + VX + W}.$$

Analytical solutions depend on $\Delta = V^2 - 4UW$ and are given in (3.56) to (3.61).

Linear K^i and linear q^i

Differential equation:

$$\frac{d\left(\overline{C_L}X\right)}{C_0 - \overline{C_L}X} = \frac{dX}{A_0X^3 + A_1X^2 + A_2X + A_3}.$$

Analytical solutions depend on

$$\Delta = A_0^2 A_3^2 - 6A_0A_1A_2A_3 + 4A_0A_2^3 + 4A_1^3A_3 - 3A_1^2A_2^2$$

and are given in (3.74), (3.76) and (3.78).

Table 3.3. Relationships among different dynamic melting models, from complex to simple models, in this chapter.

	Constant K^i	Linear K^i	Linear K^i and q^i
Incongruent Dynamic Melting	Incongruent DM w/ constant K^i	Incongruent DM w/ linear K^i	Incongruent DM w/ linear K^i and linear q^i
	$q^i = p^i \downarrow$	$q^i = p^i \downarrow$	$q^i = p^i \downarrow$
Eutectic Dynamic Melting	Eutectic DM w/ constant	Eutectic DM w/ linear K^i	Eutectic DM w/ linear K^i and linear q^i
	$x_i = x_i^0 \downarrow$	$x_i = x_i^0 \downarrow$	$x_i = x_i^0 \downarrow$
Modal Dynamic Melting	Modal DM w/ constant K^i	Modal DM w/ linear K^i	Modal DM w/ linear K^i and linear q^i

For various quantitative treatments of trace element behaviors during dynamic melting, readers can find these treatments from the literature (Maaløe, 1982; McKenzie, 1985; Williams and Gill, 1989; Sobolev and Shimizu, 1992; Albarede, 1995; Zou, 1998; Shaw, 2000; Zou, 2000; Zou and Reid, 2001).

References

Albarede, F., 1995. Introduction to Geochemical Modeling. Cambridge Univ. Press, Cambridge, 543 pp.

Maaløe, S., 1982. Geochemical aspects of permeability-controlled partial melting and fractional crystallization. Geochim. Cosmochim. Acta, 46: 43-57.

McKenzie, D., 1985. ^{230}Th- ^{238}U disequilibrium and the melting processes beneath ridge axes. Earth Planet. Sci. Lett., 72: 149-157.

Shaw, D.M., 2000. Continuous (dynamic) melting theory revisited. Canadian Mineralogist, 38: 1041-1063.

Sobolev, A.V. and Shimizu, N., 1992. Superdepleted melts and ocean mantle permeability. Doklady Rossiyskoy Akademii Nauk, 326,: 354-360.

Williams, R.W. and Gill, J.B., 1989. Effect of partial melting on the uranium decay series. Geochim. Cosmochim. Acta, 53: 1607-1619.

Zou, H.B., 1998. Trace element fractionation during modal and nonmodal dynamic melting and open-system melting: A mathematical treatment. Geochim. Cosmochim. Acta, 62: 1937-1945.

Zou, H.B., 2000. Modeling of trace element fractionation during non-modal dynamic melting with linear variations in mineral/melt distribution coefficients. Geochim. Cosmochim. Acta, 64: 1095-1102.

Zou, H.B. and Reid, M.R., 2001. Quantitative modeling of trace element fractionation during incongruent dynamic melting. Geochim. Cosmochim. Acta, 65: 153-162.

Chapter 4

Open-system Melting

Chapters 1 to 3 consider melting and melt extraction. In this chapter, we further take into account input of materials into the melting zone.

Melting in the crust and the mantle sometimes can be a open-system process (Navon and Stolper, 1987; Iwamori, 1993; O'Hara, 1995; Ozawa and Shimizu, 1995; Spiegelman, 1996; Vernieres et al., 1997; Zou, 1998; Shaw, 2000). Ozawa and Shimizu (1995) proposed a quantitative open-system melting model to explain the trace element patterns in ophiolites from a volcanic arc environment where fluid enters into the melting zone. Shaw (2000) suggested that, in addition to fluid, solid may enter into the melting zone. Besides volcanic arc environment, the addition of fluid and solid into the melting zone may take place in other tectonic settings. Melting below continents may be often related to fluid addition. In this chapter, we discuss models for open-system batch melting and open-system dynamic melting.

4.1. Open-system Batch Melting (OBM)

4.1.1. OBM with one-time instantaneous addition of melt

The one-time addition of fluids can be regarded as a one-time fluid metasomatism of the solid rocks just as it starts melting. This is a potential important model for mantle metasomatism.

The mass balance of the total mass gives

$$M_f + M_s = M_0 + M_a, \tag{4.1}$$

where M_f is the mass of the melt, M_s is the mass of the solid, M_0 the mass of the source before melting, and M_a the mass of the added melt. The balance of the mass of an element requires

$$C_f M_f + C_s M_s = C_0 M_0 + C_a M_a, \qquad (4.2)$$

where C_f is the concentration of an element in the melt, C_s is the concentration in the solid, C_0 is the concentration in the source before melting, and C_a is the concentration in the added melt.

The amount of one-time fluid metasomatism is related to the mass of the melting zone source rocks by

$$M_a = \omega M_0. \qquad (4.3)$$

Note that M_a is independent of F in this one-time addition model.

The mass of the residual solid is given by

$$M_s = (1 - F)M_0. \qquad (4.4)$$

The mass of the melt is

$$M_f = M_0 + M_{in} - M_S = (\omega + F)M_0. \qquad (4.5)$$

Substitution of (4.3), (4.4) and (4.5) into (4.2) results in

$$C_f(\omega + F)M_0 + C_S(1 - F)M_0 = C_0 M_0 + C_a \omega M_0. \qquad (4.6)$$

The concentration in the solid is related to the concentration in the melt by

$$C_S = D C_f. \qquad (4.7)$$

Combination of (4.6) and (4.7) gives

$$C_f = \frac{C_0 + C_a \omega}{(\omega + F) + D(1 - F)}. \qquad (4.8)$$

When the bulk partition coefficient is given by

$$D = \sum x_i K_i = \frac{D_0 - Q_i F}{1 - F}, \qquad (4.9)$$

we get

$$C_f = \frac{C_0 + C_a \omega}{(\omega + D_0) + (1 - Q)F}. \tag{4.10}$$

Problem. What if it is complete disequilibrium between melt and solid? For complete disequilibrium, we have

$$C_S = C_0. \tag{4.11}$$

Substitution of (4.11) into (4.6) we obtain

$$C_f(\omega + F)M_0 + C_0(1 - F)M_0 = C_0 M_0 + C_a \omega M_0. \tag{4.12}$$

Consequently, we obtain

$$C_f = \frac{C_0 F + C_a \omega}{\omega + F}. \tag{4.13}$$

4.1.2. OBM with continuous addition of melt

Fluid addition may be a one-time event or a continuous process. For the situation of continuous process, the total amount of fluids increases with the degree of melting. Thus, we can assume (Ozawa and Shimizu, 1995)

$$M_a = \beta F M_0. \tag{4.14}$$

Note that for one-time metasomatism, M_a is not related to the degree of melting.

The balance of the total mass requires

$$M_f + M_s = M_0 + M_{in}. \tag{4.15}$$

The balance of the mass for an element gives

$$C_f M_f + C_s M_s = C_0 M_0 + C_a M_a. \tag{4.16}$$

The mass of the residual solid is given by

$$M_s = (1 - F)M_0. \tag{4.17}$$

The mass of the added fluid is

$$M_a = \beta F M_0. \tag{4.18}$$

Substituting Eqs. (4.17) and (4.18) into Eq. (4.15) gives

$$M_f = M_0 + M_{in} - M_S = (1 + \beta)FM_0 . \qquad (4.19)$$

Substituting Eqs. (4.17), (4.18) and (4.19) into Eq. (4.16), we have

$$C_f(1 + \beta)FM_0 + C_S(1 - F)M_0 = C_0 M_0 + C_a \beta FM_0 . \qquad (4.20)$$

The concentration in the solid is related to that in the melt by

$$C_s = DC_f . \qquad (4.21)$$

Substitution of Eq. (4.21) into Eq. (4.20) yields

$$C_f(1 + \beta)FM_0 + DC_f(1 - F)M_0 = C_0 M_0 + C_a \beta FM_0 . \qquad (4.22)$$

Consequently,

$$C_f = \frac{C_0 + C_a \beta F}{(1 + \beta)F + D(1 - F)} . \qquad (4.23)$$

Since

$$D = \sum x_i K_i = \frac{D_0 - PF}{1 - F} , \qquad (4.24)$$

we have

$$C_f = \frac{C_0 + C_a \beta F}{D_0 + (1 + \beta - P)F} . \qquad (4.25)$$

4.1.3. OBM with continuous addition of fluids and solid

Mass balance of the total mass requires

$$M_f + M_s = M_0 + M_a + M_b . \qquad (4.26)$$

Mass balance of the trace element requires

$$C_f M_f + C_s M_s = C_0 M_0 + C_a M_a + C_b M_b . \qquad (4.27)$$

The mass of the total melt is

$$M_f = (F + \beta \gamma F)M_0 . \qquad (4.28)$$

The mass of the added melt is

$$M_a = \beta\gamma F M_0 .$$
(4.29)

The mass of added solid is

$$M_b = \beta(1-\gamma)F M_0 .$$
(4.30)

Substitution of (4.28), (4.29) and (4.30) into Eq. (4.26) yields

$$M_S = \left[1 - (1 - \beta + \beta\gamma)\right].$$
(4.31)

Substitution of (4.28), (4.29), (4.30) and (4.31) into Eq. (4.27) gives

$$
\begin{aligned}
&C_f F(1+\beta\gamma)M_0 + C_S\left[1 - F(1-\beta+\beta\gamma)\right]M_0 \\
&= C_0 M_0 + C_a \beta\gamma F_c M_0 + C_b \beta(1-\gamma)F_c M_0 .
\end{aligned}
$$
(4.32)

The concentration in the melt is related to the concentration in the solid by a bulk partition coefficient

$$C_s = DC_f .$$

Eq. (4.32) can be written as

$$
\begin{aligned}
&\left[C_0 + C_a\beta\gamma F_c + C_b\beta(1-\gamma)F_c\right]M_0 \\
&= C_f F(1+\beta\gamma)M_0 + DC_f\left[1 - F(1-\beta+\beta\gamma)\right]M_0 .
\end{aligned}
$$
(4.33)

We now need to find the variation of the bulk partition coefficient D as a function of the degree of melting. Mass balance of mineral i gives

$$M_0 x_{i,0} + M_b x_{i,0}^b = p_i L + M_s x_i .$$
(4.34)

Note that we chose L instead of $M_e + M_f$ because mineral i does not contribute to the formation of added melt. By substituting Eqs. (4.30) and (4.31) into Eq. (4.34), we have

$$
\begin{aligned}
x_i &= \frac{x_{i,0} + \beta(1-\gamma)F x_{i,0}^b - p_i F}{1-(1-\beta+\beta\gamma)F} \\
&= \frac{x_{i,0} - [p_i - \beta(1-\gamma)x_{i,0}^b]F}{1-(1-\beta+\beta\gamma)F} .
\end{aligned}
$$
(4.35)

The bulk partition coefficient is

$$D = \sum x_i K_i = \frac{D_0 - [P_i - \beta(1-\gamma)D_0^b]F}{1-(1-\beta+\beta\gamma)F} ,$$
(4.36)

where $D_0 = \sum x_{i,0} K_i$, $P_i = \sum p_i K_i$, and $D_0^b = \sum x_{i,0}^b K_i$.

From Eqs. (4.36) and (4.33) we obtain

$$C_f = \frac{C_0 + \left[C_a \gamma + C_b (1 - \gamma) \right] \beta F}{D_0 + \left[(1 - P + \beta \gamma) + \beta (1 - \gamma) D_0^b \right] F} . \qquad (4.37)$$

4.2. Open-system Dynamic Melting (ODM)

4.2.1. ODM with continuous addition of melt

Mass balance of the total mass requires

$$M_s + M_e + M_f = M_0 + M_a . \qquad (4.38)$$

Mass balance of the trace element requires

$$C_e M_e + C_f M_f + C_s M_s = C_0 M_0 + C_a M_a . \qquad (4.39)$$

Differentiation of (4.38) and (4.39) gives, respectively,

$$dM_e + dM_f + dM_s = dM_a , \qquad (4.40)$$

$$M_s dC_s + C_s dM_s + d(C_e M_e) + M_f dC_f + C_f dM_f = C_a dM_a . \qquad (4.41)$$

The only assumption so far is a constant C_a. Another piece of information arises from the exchange of the melt from the residual melt to extracted melt

$$d \left(C_e M_e \right) = C_f dM_e . \qquad (4.42)$$

Equation (4.40) can be written

$$dM_f = dM_a - dM_e - dM_s . \qquad (4.43)$$

Substitution of (4.42) and (4.43) into (4.41) yields

$$M_f dC_f + M_s dC_s = (C_f - C_s) dM_s + (C_a - C_f) dM_a . \qquad (4.44)$$

Define

$$M_s = (1 - F) M_0 , \qquad (4.45)$$

$$M_a = \beta F M_0 , \tag{4.46}$$

$$\alpha = M_f / M_s , \tag{4.47}$$

$$M_f = \alpha(1 - F)M_0 . \tag{4.48}$$

Substitution of (4.45), (4.46) and (4.48) into Eq. (4.38) results in

$$M_e = \left[(1 + \alpha + \beta)F - \alpha\right]M_0 . \tag{4.49}$$

From Eq. (4.45), we have

$$dM_s = -M_0 dF . \tag{4.50}$$

Substitution of Eq. (4.50) into Eq. (4.44) gives

$$\alpha(1-F)dC_f + (1-F)dC_s = (C_s - C_f)dF + (C_a - C_f)\beta dF . \tag{4.51}$$

The concentration in the solid is related to the concentration in the residual melt by

$$C_s = DC_f . \tag{4.52}$$

Differentiation of Eq. (4.52) using chain rule yields

$$dC_s = DdC_f + C_f dD . \tag{4.53}$$

Since

$$D = \frac{D_0 - P_i F}{1 - F} , \tag{4.54}$$

we have

$$dD = \frac{D_0 - P}{(1 - F)^2} dF . \tag{4.55}$$

Substitution of (4.54) and (4.55) into (4.53) gives

$$
\begin{aligned}
dC_s &= DdC_f + C_f dD \\
&= \frac{D_0 - PF}{1 - F} dC_f + C_f \frac{D_0 - P}{(1 - F)^2} dF .
\end{aligned}
\tag{4.56}
$$

Substitution of Eq. (4.56) into Eq. (4.51) yields

$$\frac{dC_f}{(P - 1 - \beta)C_f + C_A \beta} = \frac{dF}{(D_0 + \alpha) - (P + \alpha)F} . \tag{4.57}$$

According to Eq. (4.25), the initial condition when the melting extraction starts is

$$C_f = \frac{C_0 + C_a \beta F_C}{D_0 + (1 + \beta - P)F_C},$$ (4.58)

where F_C is the critical melting degree after which melt extraction starts. Solution to Eq. (4.57) for the concentration in the residual melt is

$$C_f = \frac{1}{P - \beta - 1} \left\{ -C_A \beta + \frac{C_A \beta D_0 + C_0 (P - \beta - 1)}{D_0 + F_c (1 + \beta - P)} \times \left[\frac{D_0 + \alpha - (\alpha + P)F}{D_0 + \alpha - (\alpha + P)F_c} \right]^{\frac{1 + \beta - P}{\alpha + p}} \right\}.$$ (4.59)

Equation (4.59) is identical to Eq. A3 in Ozawa and Shimizu (1995). The concentration in the cumulated melt is (Zou, 1998)

$$C_e = \frac{1}{F - F_C} \times \frac{1}{P - \beta - 1} \times$$

$$\left\{ -C_A \beta (F_C - F) + \frac{C_A \beta D_0 + C_0 (P - \beta - 1)}{D_0 + F_c (1 + \beta - P)} \times \frac{(D_0 + \alpha) - F_C (P + \alpha)}{1 + \alpha + \beta} \right.$$

$$\left. \times \left[1 - \left[\frac{D_0 + \alpha - (\alpha + P)F}{D_0 + \alpha - (\alpha + P)F_c} \right]^{\frac{1 + \beta - P}{\alpha + p} + 1} \right] \right\}.$$

(4.60)

4.2.2. ODM with continuous addition of melt and solid

The idea of continuous input of both melt and solid into the melting zone was first proposed by Shaw (2000) and Shaw (2006). Here we present a further full development of the model on the basis of the idea. Mass balance of the total mass requires

$$M_e + M_f + M_s = M_0 + M_a + M_b.$$ (4.61)

Mass balance of the trace element requires

$$C_e M_e + C_f M_f + C_s M_s = C_0 M_0 + C_a M_a + C_b M_b. \qquad (4.62)$$

Differentiation of Eq. (4.61) gives

$$dM_e = -d(M_f + M_s) + d(M_a + M_b). \qquad (4.63)$$

Assuming constant C_a and C_b, differentiation of Eq. (4.62) yields

$$d(C_e M_e) + d(C_f M_f) + d(C_s M_s) = C_a dM_a + C_b dM_b. \qquad (4.64)$$

Since the change of the mass of the trace element is due to the extraction of the residual melt, we have

$$d(C_e M_e) = C_f dM_e. \qquad (4.65)$$

Substitution of Eq. (4.63) into Eq. (4.65) results in

$$d(C_e M_e) = -C_f d(M_f + M_s) + C_f d(M_a + M_b). \qquad (4.66)$$

Substitution of Eq. (4.66) into Eq. (4.64) gives

$$-C_f d(M_f + M_s) + C_f d(M_a + M_b) + d(C_f M_f) + d(C_s M_s)$$
$$= C_a dM_a + C_b dM_b. $$

$$(4.67)$$

Let us define

$$M_f = \alpha M_s. \qquad (4.68)$$

The rate of addition is taken to be constant and is equal to β per unit of melting F

$$M_a + M_b = \beta F M_0. \qquad (4.69)$$

We also assume that the material added contributes to both melt and solid in mass proportions γ and $(1 - \gamma)$

$$M_a = \gamma \beta F M_0, \qquad (4.70)$$

$$M_b = (1 - \gamma)\beta F M_0. \qquad (4.71)$$

The mass of the total residual solid is

$$M_s = (1 - F)M_0 + \beta(1 - \gamma)F M_0 = \left[1 - (1 - \beta + \beta\gamma)F\right]M_0. \quad (4.72)$$

The mass of the residual melt is

$$M_f = \alpha\left[1 - (1 - \beta + \beta\gamma)F\right]M_0 . \tag{4.73}$$

The mass of the extracted melt is

$$M_e = \left\{\left[(\alpha+1)(1 - \beta + \beta\gamma)\right] - \alpha\right\}FM_0 . \tag{4.74}$$

The total mass of residual melt and extracted melt is

$$M_e + M_f = FM_0 + \beta\gamma FM_0 = (1 + \beta\gamma)FM_0 . \tag{4.75}$$

Differentiation of (4.70), (4.71), (4.72), and (4.73) with respect to F yields

$$dM_a = \gamma\beta M_0 dF , \tag{4.76}$$

$$dM_b = (1 - \gamma)\beta M_0 dF , \tag{4.77}$$

$$dM_s = -(1 - \beta + \beta\gamma)M_0 dF , \tag{4.78}$$

$$dM_f = -\alpha(1 - \beta + \beta\gamma)M_0 dF . \tag{4.79}$$

Assuming chemical equilibrium between the residual solid and the residual melt, we have

$$C_s = DC_f . \tag{4.80}$$

Mass balance of mineral i gives

$$M_0 x_{i,0} + M_b x_{i,0}^b = p_i L + M_s x_i . \tag{4.81}$$

Note that we chose L instead of $M_e + M_f$ because mineral i does not contribute to the formation of added melt.

$$x_i = \frac{x_{i,0} + \beta(1-\gamma)Fx_{i,0}^b - p_i F}{1 - (1 - \beta + \beta\gamma)F} = \frac{x_{i,0} - \left[p_i - \beta(1-\gamma)x_{i,0}^b\right]F}{1 - (1 - \beta + \beta\gamma)F} . \tag{4.82}$$

The bulk partition coefficient is

$$D = \sum x_i K_i = \frac{D_0 - \left[P_i - \beta(1-\gamma)D_0^b\right]F}{1 - (1 - \beta + \beta\gamma)F} , \tag{4.83}$$

where $D_0 = \sum x_{i,0} K_i$, $P_i = \sum p_i K_i$, and $D_0^b = \sum x_{i,0}^b K_i$. Note that
when $\beta = 0$ or $\gamma = 1$, Eq. (4.83) is reduced to the familiar expression of
Shaw (1970)

$$D = \left(D_0 - PF\right) / \left(1 - F\right).$$ (4.84)

Differentiation of Eq. (4.83) with respect to F gives

$$\frac{dD}{dF} = \frac{D_0(1 - \beta + \beta\gamma) - \left[P - \beta(1 - \gamma)D_0^b\right]}{\left[1 - (1 - \beta + \beta\gamma)F\right]^2}.$$ (4.85)

Differentiation of (4.80), we have

$$dC_s = DdC_f + C_f dD.$$ (4.86)

Since

$$D = \frac{D_0 - P_i F}{1 - F},$$ (4.87)

substitution of Eqs. (4.87) and (4.85) into Eq. (4.86), we obtain

$$dC_s = DdC_f + C_f dD$$

$$= \frac{D_0 - PF}{1 - F} dC_f + C_f \frac{D_0(1 - \beta + \beta\gamma) - \left[P - \beta(1 - \gamma)D_0^b\right]}{\left[1 - (1 - \beta + \beta\gamma)F\right]^2} dF.$$

(4.88)

Substitution of Eqs. (4.70), (4.71), (4.72), (4.73), (4.78), (4.79) and
(4.88) into (4.67) leads to

$$\frac{dC_f}{\left\{\left[P - \beta(1 - \gamma)D_0^b\right] - (1 + \beta\gamma)\right\}C_f + \beta\left[C_a\gamma + C_b(1 - \gamma)\right]}$$
$$= \frac{dF}{(\alpha + D_0) - F\left\{(1 - \beta + \beta\gamma)\alpha + \left[P - \beta(1 - \gamma)D_0^b\right]\right\}}.$$ (4.89)

The solution to Eq. (4.89) is

$$C_f = \left\{ C_0^f - \frac{[C_a\gamma + C_b(1-\gamma)]\beta}{u} \right\} \times \left[\frac{D_0 + \alpha - vF}{D_0 + \alpha - vF_c} \right]^{\frac{u}{v}}$$

$$+ \frac{[C_a\gamma + C_b(1-\gamma)]\beta}{u}, \qquad (4.90)$$

where

$$u = (1 + \beta\gamma) - \left[P - \beta(1-\gamma)D_0^b \right], \qquad (4.91)$$

$$v = (1 - \beta + \beta\gamma)\alpha + \left[P - \beta(1-\gamma)D_0^b \right]. \qquad (4.92)$$

The initial condition is given by

$$[C_0 + C_a\beta\gamma F_c + C_b\beta(1-\gamma)F_c]M_0$$
$$= C_0^f F(1 + \beta\gamma)M_0 + DC_f^0 [1 - F(1 - \beta + \beta\gamma)]M_0. \qquad (4.93)$$

And thus the first drop of the extracted melt C_0^f in Eq. (4.90) is

$$C_0^f = \frac{C_0 + [C_a\gamma + C_b(1-\gamma)]\beta F_c}{D_0 + uF_c}. \qquad (4.94)$$

It should be noted that Eq. (4.90) reduces to Eq. (4.59) when $\gamma = 1$.

4.3. Summary of Equations

1) Open-system batch melting

One-time addition of melt (Eq. (4.13)):

$$C_f = \frac{C_0 F + C_a\omega}{\omega + F}.$$

Continuous addition of melt (Eq. (4.25)):

$$C_f = \frac{C_0 + C_a\beta F}{D_0 + (1 + \beta - P)F}.$$

Continuous addition of both fluid and solid (Eq. (4.37)):

$$C_f = \frac{C_0 + \left[C_a\gamma + C_b(1-\gamma)\right]\beta F}{D_0 + \left[(1-P+\beta\gamma) + \beta(1-\gamma)D_0^b\right]F}.$$

2) Open-system dynamic melting

Continuous addition of melt (Eq. (4.59)):

$$C_f = \frac{1}{P-\beta-1}\left\{ -C_A\beta + \frac{C_A\beta D_0 + C_0(P-\beta-1)}{D_0 + F_c(1+\beta-P)} \times \left[\frac{D_0 + \alpha - (\alpha+P)F}{D_0 + \alpha - (\alpha+P)F_c} \right]^{\frac{1+\beta-P}{\alpha+p}} \right\}.$$

Continuous addition of both melt and solid (Eq. (4.90)):

$$C_f = \left\{ C_0^f - \frac{\left[C_a\gamma + C_b(1-\gamma)\right]\beta}{u} \right\} \times \left[\frac{D_0 + \alpha - vF}{D_0 + \alpha - vF_c} \right]^{\frac{u}{v}}$$
$$+ \frac{\left[C_a\gamma + C_b(1-\gamma)\right]\beta}{u}.$$

where

$$C_0^f = \frac{C_0 + \left[C_a\gamma + C_b(1-\gamma)\right]\beta F_c}{D_0 + uF_c}.$$

References

Iwamori, H., 1993. Dynamic disequilibrium melting model with porous flow and diffusion-controlled chemical equilibration. Earth Planet. Sci. Lett., 114: 301-313.

Navon, O. and Stolper, E., 1987. Geochemical consequences of melt percolation: The upper mantle as a chromatographic column. J. Geology, 95: 285-307.

O'Hara, M.J., 1995. Imperfect melting separation, finite increment size and source region flow during fractional melting and the generation of reversed or dubdued discrimination of incompatible trace elements. Chem. Geol., 121: 27-50.

Ozawa, K. and Shimizu, N., 1995. Open-system melting in the upper mantle: Constraints from the Hayachine-Miyamori ophiolite, northeastern Japan. J. Geophys. Res., 100: 22315-22335.

Shaw, D.M., 2000. Continuous (dynamic) melting theory revisited. Canadian Mineralogist, 38(5): 1041-1063.

Shaw, D.M., 2006. Trace Elements in Magmas: A Theoretical Treatment. Cambridge University Press, Cambridge, 243 pp.

Spiegelman, M., 1996. Geochemical consequences of melt transport in 2-D: The sensitivity of trace elements to mantle dynamics. Earth Planet. Sci. Lett., 139: 115-132.

Vernieres, J., Godard, M. and Bodinier, J.L., 1997. A plate model for the simulation of trace element fractionation during partial melting and magma transport in the Earth's upper mantle. J. Geophys. Res., 102: 24771-24784.

Zou, H.B., 1998. Trace element fractionation during modal and nonmodal dynamic melting and open-system melting: A mathematical treatment. Geochim. Cosmochim. Acta, 62: 1937-1945.

Chapter 5

Uranium-series Disequilibrium Modeling

The previous four chapters deal with the fractionation of stable trace elements during partial melting. In this chapter, we study the behaviors of radioactive uranium decay series during partial melting. Since quantitative models for uranium-series disequilibria need to include additional parameters in decay constants and are thus more complicated, for simplicity, we assume that the partition coefficients remain constant during partial melting. Thus, we only present modal dynamic melting.

^{238}U decays to stable ^{206}Pb, and ^{235}U decays to stable ^{207}Pb via two different chains of short-lived intermediate nuclides with a wide range of half-lives. Among these intermediate nuclides, ^{230}Th and ^{226}Ra from ^{238}U and ^{231}Pa from ^{235}U are of particular importance to earth sciences. The half-lives of ^{230}Th, ^{226}Ra and ^{231}Pa are 75,200 y, 1600 y, and 32,800 y, respectively, and bracket the time scales of geological processes.

Before we work on the uranium decay series, we start with single-stage radioisotope decay.

5.1. Single-stage Radioisotope Decay

For the decay of N number of radioactive atoms, the rate of decay, dN/dt, is proportional to the number of atoms N,

$$\frac{dN}{dt} = -\lambda N, \qquad (5.1)$$

where λ, the decay constant, is essentially the probability of decay per unit time.

The solution to Eq. (5.1) with initial condition $N(0) = N_0$ is

$$N = N_0 e^{-\lambda t}. \qquad (5.2)$$

The number of radiogenic daughter atoms formed, G^*, equals the number of parent atoms consumed:

$$G^* = N_0 - N. \qquad (5.3)$$

Equation (5.2) can be re-expressed as

$$N_0 = N e^{\lambda t}. \qquad (5.4)$$

Combination of Eq. (5.3) and Eq. (5.4) yields

$$G^* = N(e^{\lambda t} - 1).$$

A useful way of referring to the rate of decay of a radionuclide is the half-life, $t_{1/2}$, which is the time required for half of the parent atoms to decay. And according to Eq. (5.2), when $N = N_0/2$, we obtain

$$t_{1/2} = \frac{\ln 2}{\lambda} = \frac{0.693}{\lambda}. \qquad (5.5)$$

If the number of daughter atoms at time $t = 0$ is G_0, then the total number of atoms at time t is

$$G = G_0 + G^* = G_0 + N(e^{\lambda t} - 1). \qquad (5.6)$$

Dividing Eq. (5.6) by the number of atoms of a stable daughter isotope (G_B), we obtain

$$\frac{G}{G_B} = \frac{G_0}{G_B} + \frac{N}{G_B}(e^{\lambda t} - 1). \qquad (5.7)$$

Equation (5.7), rather than (5.6), is used in isotope geochemistry, because mass spectrometers can measure isotopic ratios (G/G_B) significantly more precise than the amount of an isotope (G).

Equation (5.7) is the fundamental equation of single-stage decay in geochronology. For example, from Eq. (5.7), for the ^{87}Rb-^{87}Sr system,

$$\frac{^{87}Sr}{^{86}Sr} = \left(\frac{^{87}Sr}{^{86}Sr}\right)_0 + \frac{^{87}Rb}{^{86}Sr}(e^{\lambda_{87Rb}t} - 1); \qquad (5.8)$$

for the ^{147}Sm-^{143}Nd system,

$$\frac{^{143}Nd}{^{144}Nd} = \left(\frac{^{143}Nd}{^{144}Nd}\right)_0 + \frac{^{147}Sm}{^{144}Nd}(e^{\lambda_{147Sm}t} - 1) ; \qquad (5.9)$$

for the ^{176}Lu-^{176}Hf system,

$$\frac{^{176}Hf}{^{177}Hf} = \left(\frac{^{176}Hf}{^{177}Hf}\right)_0 + \frac{^{176}Lu}{^{177}Hf}(e^{\lambda_{176Lu}t} - 1) ; \qquad (5.10)$$

and for the ^{187}Re-^{187}Os system,

$$\frac{^{187}Os}{^{188}Os} = \left(\frac{^{187}Os}{^{188}Os}\right)_0 + \frac{^{187}Re}{^{188}Os}(e^{\lambda_{187Re}t} - 1) . \qquad (5.11)$$

The decays of ^{238}U, ^{235}U and ^{232}Th are more complicated. If we ignore the intermediate isotopes in the U and Th decay series, then for the ^{238}U-^{206}Pb system

$$\frac{^{206}Pb}{^{204}Pb} = \left(\frac{^{206}Pb}{^{204}Pb}\right)_0 + \frac{^{238}U}{^{204}Pb}(e^{\lambda_{238U}t} - 1) ; \qquad (5.12)$$

for the ^{235}U-^{207}Pb system

$$\frac{^{207}Pb}{^{204}Pb} = \left(\frac{^{207}Pb}{^{204}Pb}\right)_0 + \frac{^{235}U}{^{204}Pb}(e^{\lambda_{235U}t} - 1) ; \qquad (5.13)$$

and for the ^{232}Th-^{208}Pb system

$$\frac{^{208}Pb}{^{204}Pb} = \left(\frac{^{208}Pb}{^{204}Pb}\right)_0 + \frac{^{232}Th}{^{204}Pb}(e^{\lambda_{232Th}t} - 1) . \qquad (5.14)$$

However, for investigating short -term recent geological processes, the intermediate isotopes in decay series cannot be ignored and in fact these short-lived intermediate isotopes are powerful in deciphering young geological processes.

5.2. Decay Series

Before we study the behaviors of U-series disequilibria during partial melting, we try to understand a simple case of U-decay series in an isolated system in the absence of melting. For this undisturbed closed system, the uranium-thorium-radium (U-Th-Ra) series disequilibria (Fig.

5.1) can be described by the following system of three differential equations:

$$\frac{d^{238}U}{dt} = -\lambda_{238}{}^{238}U,$$ (5.15)

$$\frac{d^{230}Th}{dt} = \lambda_{238}{}^{238}U - \lambda_{230}{}^{230}Th,$$ (5.16)

$$\frac{d^{226}Ra}{dt} = \lambda_{230}{}^{230}Th - \lambda_{226}{}^{226}Ra.$$ (5.17)

where λ_{238}=1.55125×10^{-10} year^{-1}, λ_{230} =9.217×10^{-6} year^{-1} and λ_{226} =4.272×10^{-4} year^{-1} are decay constants of ^{238}U, ^{230}Th and ^{226}Ra, respectively.

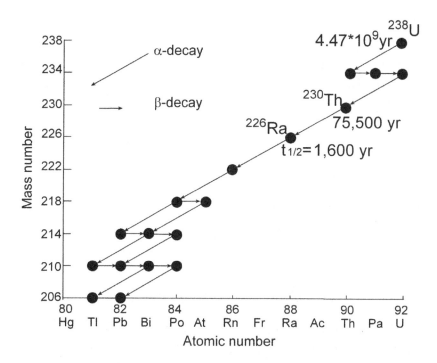

Fig. 5.1. Schematics showing the decay series of ^{238}U to the final stable ^{206}Pb. Among the intermediate isotopes, ^{230}Th and ^{226}Ra have wide applications in geological and environmental studies. $t_{1/2}$ denotes half life.

The exact solutions to the above system of differential equation (5.15), (5.16) and (5.17) are

$$^{238}U = {}^{238}U_0 e^{-\lambda_{238}t},$$ (5.18)

$$^{230}Th = \frac{\lambda_{238}}{\lambda_{230} - \lambda_{238}} \,^{238}U_0 \left(e^{-\lambda_{238}t} - e^{-\lambda_{230}t} \right) + {}^{230}Th_0 e^{-\lambda_{230}t},$$ (5.19)

$$^{226}Ra = {}^{226}Ra_0 \, e^{-\lambda_{226}t}$$

$$+ \frac{\lambda_{230}}{\lambda_{226} - \lambda_{230}} \,^{230}Th_0 \left[e^{-\lambda_{230}t} - e^{-\lambda_{226}t} \right]$$ (5.20)

$$+ \frac{\lambda_{230}\lambda_{238}}{\lambda_{230} - \lambda_{238}} \,^{238}U_0 \left[\frac{e^{-\lambda_{238}t} - e^{-\lambda_{226}t}}{\lambda_{226} - \lambda_{238}} - \frac{e^{-\lambda_{230}t} - e^{-\lambda_{226}t}}{\lambda_{226} - \lambda_{230}} \right].$$

5.2.1. Bateman equation

Bateman's solutions to decay series were derived by assuming that there are no initial daughter nuclides in the system. For this case, the assumptions are

$$^{230}Th_0 = 0,$$

and

$$^{226}Ra_0 = 0.$$

Then Eqs. (5.19) and (5.20) reduce to

$$^{230}Th = \frac{\lambda_{238}}{\lambda_{230} - \lambda_{238}} \,^{238}U_0 \left(e^{-\lambda_{238}t} - e^{-\lambda_{230}t} \right),$$ (5.21)

$$^{226}Ra = \frac{\lambda_{230}\lambda_{238}}{\lambda_{230} - \lambda_{238}} \,^{238}U_0 \left[\frac{e^{-\lambda_{238}t} - e^{-\lambda_{226}t}}{\lambda_{226} - \lambda_{238}} - \frac{e^{-\lambda_{230}t} - e^{-\lambda_{226}t}}{\lambda_{226} - \lambda_{230}} \right],$$ (5.22)

respectively. Bateman's solutions (5.21) and (5.22) are useful for the scenario when no initial daughter nuclides are present. Equations (5.21) and (5.22), however, cannot be applied to the uranium-series studies of volcanic rocks, because the initial $^{230}Th_0$ and $^{226}Ra_0$ in volcanic rocks are not equal to zero.

5.2.2. Secular equilibrium

Since λ_{238} is significantly smaller than λ_{230} and λ_{226}, as an approximation, we have $\lambda_{230} - \lambda_{238} \approx \lambda_{230}$, and $\lambda_{226} - \lambda_{238} \approx \lambda_{226}$. If t is large enough, we have

$$e^{-\lambda_{230}t} \approx 0,$$

$$e^{-\lambda_{226}t} \approx 0.$$

Thus, for large t, by ignoring all terms containing $e^{-\lambda_{230}t}$ and $e^{-\lambda_{226}t}$, Eq. (5.19) for ^{230}Th and Eq. (5.20) for ^{226}Ra can be simplified as

$$^{230}Th = \frac{\lambda_{238}}{\lambda_{230}} \, {}^{238}U_0 e^{-\lambda_{238}t} = \frac{\lambda_{238}}{\lambda_{230}} \, {}^{238}U, \qquad (5.23)$$

$$^{226}Ra = \frac{\lambda_{238}}{\lambda_{226}} \, {}^{238}U_0 e^{-\lambda_{238}t} = \frac{\lambda_{238}}{\lambda_{226}} \, {}^{238}U, \qquad (5.24)$$

respectively.
From Eq. (5.23) we have

$$\lambda_{230} \, {}^{230}Th = \lambda_{238} \, {}^{238}U.$$

From Eq. (5.24) we get

$$\lambda_{226} \, {}^{226}Ra = \lambda_{238} \, {}^{238}U.$$

Consequently, we obtain

$$\lambda_{238} \, {}^{238}U = \lambda_{230} \, {}^{230}Th = \lambda_{226} \, {}^{226}Ra. \qquad (5.25)$$

The system that satisfies Eq. (5.25) is in secular equilibrium. Since the activity denoted by parenthesis is defined as the product of decay constant and the amount of radioactive nuclides,

$$(N) = \lambda N, \qquad (5.26)$$

for uranium decay series, according to Eqs. (5.25) and (5.26), all nuclides in a decay series have the same activities when the secular equilibrium is reached:

$$(^{238}U) = (^{230}Th) = (^{226}Ra). \qquad (5.27)$$

Only in this chapter, parenthesis represents activity.

Thus the sufficient condition to reach secular equilibrium in an undisturbed geological system is enough isolated time for the system. The time to reach secular equilibrium is normally five half-lives of the daughter nuclides. For ^{230}Th, the half-life for is 75,000 years, and after 375,000 years, or five half lives of ^{230}Th, the system reaches ^{238}U-^{230}Th equilibrium. Similarly, the half-life for ^{226}Ra is 1,600 years, and Th-Ra secular equilibrium can be reached after five half-lives of ^{226}Ra, or 8,000 years.

If there is no initial ^{230}Th in the system, that is, $^{230}Th_0 = 0$, using the activity notation, $(^{230}Th) = \lambda_{230}\,^{230}Th$ and $(^{238}U) = \lambda_{238}\,^{238}U$, then from the Bateman equations (5.21) and (5.18), we obtain

$$\left(\frac{^{230}Th}{^{238}U} \right) = \frac{\lambda_{230}}{\lambda_{230} - \lambda_{238}} \left[1 - e^{(\lambda_{238} - \lambda_{230})t} \right]. \tag{5.28}$$

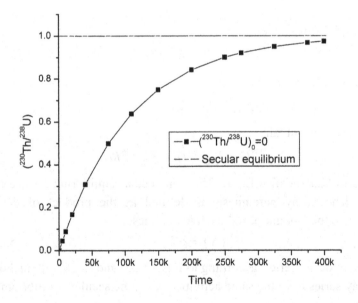

Fig. 5.2. Plot of Eq. (5.28) for the evolution of $(^{230}$Th/^{238}U) with time when initial $(^{230}$Th) is zero. The horizontal secular equilibrium line is also shown for comparison. $k = 1000$ years. It can be seen that secular ^{230}Th-^{238}U equilibrium can be reached in about 400 thousand years.

Since both λ_{238} and λ_{230} are known, $(^{230}\text{Th}/^{238}\text{U})$ from Eq. (5.28) is a function of time only. Figure 5.2 displays the growth of ^{230}Th with time when initial $(^{230}Th/^{238}U)_0 = 0$.

If there are initial ^{230}Th nuclides in the system, using the activity notation, from Eqs. (5.18) and (5.19), we can obtain the equation for $(^{230}\text{Th}/^{238}\text{U})$ with initial ^{230}Th:

$$\left(\frac{^{230}Th}{^{238}U}\right) = \frac{\lambda_{230}}{\lambda_{230} - \lambda_{238}}\left[1 - e^{(\lambda_{238}-\lambda_{230})t}\right] + \left(\frac{^{230}Th}{^{238}U}\right)_0 e^{(\lambda_{238}-\lambda_{230})t} \quad (5.29)$$

$(^{230}\text{Th}/^{238}\text{U})$ from Eq. (5.29) is a function of time and the initial activity ratio $(^{230}Th/^{238}U)_0$. Equation (5.29) reduces to Eq. (5.28) if $(^{230}Th/^{238}U)_0 = 0$. It is noted that $(^{230}Th/^{238}U)_0$ may be greater than or less than unity. When $(^{230}Th/^{238}U)_0 > 1.0$, it is originally ^{230}Th excess; if $(^{230}Th/^{238}U)_0 < 1.0$, it is originally ^{238}U excess. Figure 5.2 shows that secular equilibrium recovers in 375 thousand years for different initial $(^{230}Th/^{238}U)_0$ values.

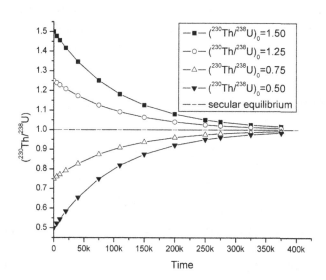

Fig. 5.2. Evolution of $(^{230}\text{Th}/^{238}\text{U})$ with time for different initial $(^{230}\text{Th}/^{238}\text{U})_0$ ratios. After 375k (375,000) years, the initial ^{230}Th or ^{238}U excess disappears and the system returns to secular equilibrium with $(^{230}\text{Th}/^{238}\text{U})=1.0$.

5.3. U-series Disequilibrium Produced by Dynamic Melting

The secular equilibrium may not last forever, and can be disturbed by partial melting. Since the half-lives of the intermediate nuclides ([230]Th, [226]Ra, and [231]Pa) are short relative to the time scale of mantle evolution, these nuclides can be assumed to be in secular equilibrium at the initiation of melting. In addition, mantle sources can be safely assumed to be in radioactive equilibrium regardless of their chemical compositions, which avoids uncertainties in initial source compositions before melting. Therefore, observed secular disequilibria among the uranium-series (U-series) nuclides in young lavas may provide important information on mantle melting processes and the time scales of melt generation.

We will present quantitative models that relate the extents of U-series disequilibria to the melting process as a function of elemental partition coefficients, melting porosity, melting rate and melting time. We then quantitatively investigate the effect of ϕ, $\dot{M}$, T and the distribution coefficients of U and Th on U-series disequilibria in young lavas, and quantify the relative contributions from net elemental fractionation and ingrowth of daughter nuclides to the total U-series disequilibria.

If solid and melt are in equilibrium during partial melting, the conservation law which is concerned with the concentration of a radioactive and radiogenic nuclide in the solid and melt is given by McKenzie (1985) as the following differential equation

$$\left[\rho_f \phi + \rho_s \left(1-\phi\right) D_d\right]\frac{\partial C_f^d}{\partial t} = \left(D_d - 1\right) C_f^d \dot{M}$$

$$-\lambda_d \left[\rho_f \phi + \rho_s \left(1-\phi\right) D_d\right] C_f^d \qquad (5.30)$$

$$+\lambda_p \left[\rho_f \phi + \rho_s \left(1-\phi\right) D_p\right] C_f^p.$$

The first term on the right-hand side is the melting term; the second term on the right-hand side is the radioactive decay of the nuclide; and the last term on the right-hand side represents the radiogenic production by the parent of the nuclide. C_f^d and C_f^p are the concentrations of the nuclide and its parent in the melt, respectively. D_d and D_p are the distribution coefficients of the nuclide and its parent, respectively. ρ_s is the density of the solid (3300 kg/m^3), ρ_f is the density of the melt (2800 kg/m^3), ϕ is the volume porosity of the mantle, $\dot{M}$ is the melting rate (kg/m^3/year), λ_d and λ_p are the decay constants of the daughter nuclide and its parent, respectively.

The concentration of the daughter nuclide in the solid C_s^d is related to the daughter concentration in the melt C_f^d by

$$C_s^d = D_d C_f^d \, ,$$

and the concentration of the parent nuclide in the solid C_s^p is related to the concentration of the parent in the melt C_f^p by

$$C_s^p = D_p C_f^p \, .$$

Equation (5.30) can be rewritten as

$$\frac{\partial \left(C_f^d \right)}{\partial t} = -\left(\alpha_d + \lambda_d \right)\left(C_f^d \right) + \lambda_d \frac{F_d}{F_p} \left(C_f^p \right), \tag{5.31}$$

where

$$\alpha_d \left(\phi, \dot{M} \right) = \frac{\left(1 - D_d \right) \dot{M}}{\rho_f \phi + D_d \rho_s \left(1 - \phi \right)}, \tag{5.32}$$

$$F_i \left(\phi \right) = \frac{\rho_f \phi}{D_i \rho_s \left(1 - \phi \right) + \rho_f \phi}, \tag{5.33}$$

and $\left(C_f^d \right)$ and $\left(C_f^p \right)$ are the activities [defined as the product of its concentration C and decay constant λ, and denoted by parentheses, $\left(C_f^d \right) = \lambda_d C_f^d$, and $\left(C_f^p \right) = \lambda_p C_f^p$] of the daughter nuclide and the parent nuclide in the melt, respectively. α_d is the melting parameter for the nuclide and has the same dimension (year^{-1}) as the radioactive decay constant λ. If there is no melting, that is, the melting rate $\dot{M} = 0$, from Eq. (5.32) we have $\alpha = 0$. The subscript i in Eq. (5.33) may represent a daughter or its parent.

Equation (5.31) can be used to model the uranium decay series (U-series) nuclides in the residual melt that is in chemical equilibrium with the solid during dynamic partial melting:

$$\frac{d\left(^{238}U\right)_f}{dt} = -\left(\alpha_U + \lambda_{238}\right)\left(^{238}U\right)_f, \tag{5.34}$$

$$\frac{d\left(^{230}Th\right)_f}{dt} = -\left(\alpha_{Th} + \lambda_{230}\right)\left(^{230}Th\right)_f$$
$$+\lambda_{230}\frac{F_{Th}}{F_U}\left(^{238}U\right)_f, \tag{5.35}$$

$$\frac{d\left(^{226}Ra\right)_f}{dt} = -\left(\alpha_{Ra} + \lambda_{226}\right)\left(^{226}Ra\right)_f$$
$$+\lambda_{226}\frac{F_{Ra}}{F_{Th}}\left(^{230}Th\right)_f. \tag{5.36}$$

$\left(^{238}U\right)_f$, $\left(^{230}Th\right)_f$, and $\left(^{226}Ra\right)_f$ are the activities of ^{238}U, ^{230}Th, and ^{226}Ra in the residual melt. α_U, α_{Th} and α_{Ra} are given by Eq. (5.32) and F_U, F_U and F_U are calculated by Eq. (5.33).

Equation (5.34) is different from Eq. (5.35) and Eq. (5.36) because ^{238}U, unlike ^{230}Th or ^{226}Ra, has no parent. When melting rate ($\dot{M}$) is constant in the upwelling material and the melt fraction (ϕ) in equilibrium with the matrix remains constant, both α and F are independent of time (see Eqs. (5.32) and (5.33)) and the system of differential equations (5.34), (5.35) and (5.36) has explicit solutions:

$$\left(^{238}U\right)_f = \left(^{238}U\right)_f^0 \exp\left[-\left(\alpha_U + \lambda_{238}\right)t\right], \tag{5.37}$$

$$\left(^{230}Th\right)_f = \left(^{230}Th\right)_f^0 \exp\left[-\left(\alpha_{Th} + \lambda_{230}\right)t\right]$$
$$+\frac{\lambda_{230}}{\alpha_{Th} + \lambda_{230} - \alpha_U - \lambda_{238}}\frac{F_{Th}}{F_U}\left(^{238}U\right)_f^0 \times \tag{5.38}$$
$$\left\{\exp\left[-\left(\alpha_U + \lambda_{238}\right)t\right] - \exp\left[\left(\alpha_{Th} + \lambda_{230}\right)t\right]\right\},$$

$$
\left(^{226}Ra\right)_f = \left(^{226}Ra\right)_f^0 \exp\left[-(\alpha_{Ra} + \lambda_{226})t\right]
$$

$$
+ \frac{\lambda_{226}}{\alpha_{Ra} + \lambda_{226} - \alpha_{Th} - \lambda_{230}} \frac{F_{Ra}}{F_{Th}} \left(^{230}Th\right)_f^0 \times
$$

$$
\left\{\exp\left[-(\alpha_{Th} + \lambda_{230})t\right] - \exp\left[-(\alpha_{Ra} + \lambda_{226})t\right]\right\}
$$

$$
+ \frac{\lambda_{230}\lambda_{226}}{\alpha_{Th} + \lambda_{230} - \alpha_{U} - \lambda_{238}} \frac{F_{Ra}}{F_{U}} \left(^{238}U\right)_f^0 \times
$$

$$
\left\{
\begin{array}{c}
\dfrac{\exp\left[-(\alpha_{U} + \lambda_{238})t\right] - \exp\left[-(\alpha_{Ra} + \lambda_{226})t\right]}{\alpha_{Ra} + \lambda_{226} - \alpha_{U} - \lambda_{238}} \\[3mm]
- \dfrac{\exp\left[-(\alpha_{Th} + \lambda_{230})t\right] - \exp\left[-(\alpha_{Ra} + \lambda_{226})t\right]}{\alpha_{Ra} + \lambda_{226} - \alpha_{Th} - \lambda_{230}}
\end{array}
\right\},
$$

$$
(5.39)
$$

Note that solutions (5.37), (5.38) and (5.39) collapse to Eqs. (5.18), (5.19) and (5.20), respectively, when

$$
\alpha_{U} = \alpha_{Th} = \alpha_{Ra} = 0,
$$

and

$$
F_{U} = F_{Th} = F_{Ra}.
$$

The above conditions are satisfied when there is not melting in the closed system where melting rate $\dot{M} = 0$ and there is no elemental partitioning.

The solutions for the residual melt are the same as those obtained by (McKenzie, 1985) for $\left(^{238}U\right)_f$ and $\left(^{230}Th\right)_f$, and (Williams and Gill, 1989) for $\left(^{226}Ra\right)_f$ except that the decay constant of ^{238}U (λ_{238}) is included here. As basalts are considered as extracted melts, the next important step is to derive the equations for the extracted melt from the solutions for the residual melt. The real difference of this study from previous ones starts from this step. The ^{238}U, ^{230}Th and ^{226}Ra activities of the extracted melt produced by dynamic melting will be average values of the residual melt. An average activity of a nuclide in the extracted melt, (C), can be approximately obtained by averaging relative to melting time (T) (McKenzie, 1985; Chabaux and J., 1994),

$$(C) = \frac{1}{T} \int_0^T \left(C_f(t)\right) dt. \tag{5.40}$$

The accurate activity of the nuclide in the extracted melt will be the average values of these functions for the residual melt with respect to X, the mass fraction of extracted melt relative to the initial amount (Williams and Gill, 1989)

$$(C) = \frac{1}{X} \int_0^X \left(C_f(X)\right) dX = \frac{1}{X} \int_0^T \left(C_f(t) \frac{dX}{dt}\right) dt. \tag{5.41}$$

Another key for the derivation of the equation for (C) is to accurately express X as a function of ϕ, $\dot{M}$, and t in order to obtain dX/dt in Eq. (5.41). X is related to $\dot{M}_e$ (the melt extraction rate), ϕ, and t by the following relationship (Zou, 1998)

$$X = 1 - \exp\left[-\frac{\dot{M}_e}{\rho_f \phi + \rho_s (1-\phi)} t\right], \tag{5.42}$$

where $\dot{M}_e$ is related to the melting rate $\dot{M}$ by (Zou, 1998)

$$\frac{\dot{M}}{\dot{M}_e} = \frac{\rho_s (1-\phi)}{\rho_f \phi + \rho_s (1-\phi)}. \tag{5.43}$$

Combination of Eq. (5.42) with Eq. (5.43) yields

$$X(\phi, \dot{M}, t) = 1 - \exp\left[-\frac{\dot{M}}{\rho_s (1-\phi)} t\right]. \tag{5.44}$$

Differentiation of Eq. (5.44) leads to

$$\frac{dX}{dt} = \frac{\dot{M}}{\rho_s (1-\phi)} \exp\left[-\frac{\dot{M}}{\rho_s (1-\phi)} t\right]. \tag{5.45}$$

By combining Eq. (5.41), Eq.   with Eqs. (5.37), (5.38), (5.39), respectively, we obtain

$$\frac{\left(^{238}U\right)X}{\theta} = \frac{\left(^{238}U\right)_f^0}{\alpha_U + \lambda_{238} + \theta}\left\{1 - \exp\left[-(\alpha_U + \lambda_{238} + \theta)T\right]\right\}. \tag{5.46}$$

$$\frac{\left(^{230}Th\right)X}{\theta} = \frac{\lambda_{230}}{\alpha_{Th} + \lambda_{230} - \alpha_U - \lambda_{238}} \frac{1}{\alpha_U + \lambda_{238} + \theta} \frac{F_{Th}}{F_U} \times$$

$$\left(^{238}U\right)_f^0 \left\{1 - \exp\left[-(\alpha_U + \lambda_{238} + \theta)T\right]\right\}$$

$$+ \frac{\left(\alpha_{Th} + \lambda_{230} - \alpha_U - \lambda_{238}\right)\left(^{230}Th\right)_f^0 - \lambda_{230}\left(\frac{F_{Th}}{F_U}\right)\left(^{238}U\right)_f^0}{\left(\alpha_{Th} + \lambda_{230} + \theta\right)\left(\alpha_{Th} + \lambda_{230} - \alpha_U - \lambda_{238}\right)} \times$$

$$\left\{1 - \exp\left[-(\alpha_{Th} + \lambda_{230} + \theta)T\right]\right\},$$

$$(5.47)$$

$$\frac{\left(^{226}Ra\right)X}{\theta} = \frac{\lambda_{230}\lambda_{226}}{\left(\alpha_{Th} + \lambda_{230} - \alpha_U - \lambda_{238}\right)\left(\alpha_{Ra} + \lambda_{226} - \alpha_U - \lambda_{238}\right)} \frac{F_{Ra}}{F_U} \times$$

$$\left(^{238}U\right)_f^0 \frac{1 - \exp\left[-(\alpha_U + \lambda_{238} + \theta)T\right]}{\alpha_U + \lambda_{238} + \theta}$$

$$+ \frac{1}{\alpha_{Ra} + \lambda_{226} - \alpha_{Th} - \lambda_{230}} \times$$

$$\left[\lambda_{226}\frac{F_{Ra}}{F_{Th}}\left(^{230}Th\right)_f^0 - \frac{\lambda_{230}\lambda_{226}}{\alpha_{Th} + \lambda_{230} - \alpha_U - \lambda_{238}}\frac{F_{Ra}}{F_U}\left(^{238}U\right)_f^0\right] \times$$

$$\frac{1 - \exp\left[-(\alpha_{Th} + \lambda_{230} + \theta)T\right]}{\alpha_{Th} + \lambda_{230} + \theta}$$

$$+ \left[\left(^{226}Ra\right)_f^0 - \frac{\lambda_{226}}{\alpha_{Ra} + \lambda_{226} - \alpha_{Th} - \lambda_{230}}\frac{F_{Ra}}{F_{Th}}\left(^{230}Th\right)_f^0 + \frac{\left(^{238}U\right)_f^0 \lambda_{230}\lambda_{226}F_{Ra}/F_U}{\left(\alpha_{Ra} + \lambda_{226} - \alpha_{Th} - \lambda_{230}\right)\left(\alpha_{Ra} + \lambda_{226} - \alpha_U - \lambda_{238}\right)}\right] \times$$

$$\frac{1 - \exp\left[-(\alpha_{Ra} + \lambda_{226} + \theta)T\right]}{\alpha_{Ra} + \lambda_{226} + \theta},$$

$$(5.48)$$

where

$$\theta = \frac{\dot{M}}{\rho_s (1-\phi)}, \tag{5.49}$$

is a function of ϕ and $\dot{M}$. $\left(^{238}U\right)_f^0$, $\left(^{230}Th\right)_f^0$, and $\left(^{226}Ra\right)_f^0$ are the initial activities of ^{238}U, ^{230}Th, and ^{226}Ra in the first drop of the extracted melt. It should be emphasized that, when setting all the decay constants to be zero, Eqs. (5.46), (5.47), (5.48) reduce to Eq. 14 in Zou (1998) which describes trace element concentrations in the extracted melt in the context of modal dynamic melting

$$C = \frac{1}{X} C_o \left\{ 1 - (1-X)^{\left[\rho_f \phi + \rho_s (1-\phi)\right]/\left[\rho_f \phi + \rho_s (1-\phi)D\right]} \right\}. \tag{5.50}$$

The final step is to obtain ready-to-use equations for $\left(^{230}Th/^{238}U\right)$ and $\left(^{226}Ra/^{230}Th\right)$. To do so, we need to know the relationship among the initial conditions $\left(^{238}U\right)_f^0$, $\left(^{230}Th\right)_f^0$, and $\left(^{226}Ra\right)_f^0$ in Eqs. (5.46), (5.47), (5.48) so as to eventually eliminate these initial conditions from the expressions for $\left(^{230}Th/^{238}U\right)$ and $\left(^{226}Ra/^{230}Th\right)$. The relationship between $\left(^{230}Th\right)_f^0$ and $\left(^{238}U\right)_f^0$ is (O'Nions and McKenzie, 1993)

$$\left(^{230}Th\right)_f^0 \Big/ \left(^{238}U\right)_f^0 = F_{Th}/F_U. \tag{5.51}$$

Similarly, we have

$$\left(^{226}Ra\right)_f^0 \Big/ \left(^{230}Th\right)_f^0 = F_{Ra}/F_{Th}, \tag{5.52}$$

$$\left(^{231}Pa\right)_f^0 \Big/ \left(^{235}U\right)_f^0 = F_{Pa}/F_U. \tag{5.53}$$

Substituting Eqs. (5.51), (5.52), (5.53) into Eqs. (5.47) and (5.48), respectively, we find

$$\left(^{230}Th\right) = \frac{\left(^{230}Th\right)_f^0 \theta}{X(\alpha_U + \lambda_{238} + \theta)} \left\{ \begin{array}{l} c_1 \left\{ 1 - \exp\left[-(\alpha_U + \lambda_{238} + \theta)T \right] \right\} + \\ c_2 \left\{ 1 - \exp\left[-(\alpha_{Th} + \lambda_{230} + \theta)T \right] \right\} \end{array} \right\},$$

$$\tag{5.54}$$

$$\left(^{226}Ra\right) = \frac{\left(^{226}Ra\right)_f^0 \theta}{X\left(\alpha_U + \lambda_{238} + \theta\right)} \begin{cases} c_3\left\{1 - \exp\left[-\left(\alpha_U + \lambda_{238} + \theta\right)T\right]\right\} + \\ c_4\left\{1 - \exp\left[-\left(\alpha_{Th} + \lambda_{230} + \theta\right)T\right]\right\} + \\ c_5\left\{1 - \exp\left[-\left(\alpha_{Ra} + \lambda_{226} + \theta\right)T\right]\right\} \end{cases},$$

(5.55)

where

$$c_1 = \frac{\lambda_{230}}{\alpha_{Th} + \lambda_{230} - \alpha_U - \lambda_{238}},$$

(5.56)

$$c_2 = \frac{\left(\alpha_U + \lambda_{238} + \theta\right)\left(\alpha_{Th} - \alpha_U - \lambda_{238}\right)}{\left(\alpha_{Th} + \lambda_{230} + \theta\right)\left(\alpha_{Th} + \lambda_{230} - \alpha_U - \lambda_{238}\right)},$$

(5.57)

$$c_3 = \frac{\lambda_{230}\lambda_{226}}{\left(\alpha_{Th} + \lambda_{230} - \alpha_U - \lambda_{238}\right)\left(\alpha_{Ra} + \lambda_{226} - \alpha_U - \lambda_{238}\right)},$$

(5.58)

$$c_4 = \frac{\lambda_{226}\left(\alpha_U + \lambda_{238} + \theta\right)}{\left(\alpha_{Th} + \lambda_{230} + \theta\right)\left(\alpha_{Ra} + \lambda_{226} - \alpha_{Th} - \lambda_{230}\right)} \times$$
$$\frac{\alpha_{Th} - \alpha_U - \lambda_{238}}{\alpha_{Th} + \lambda_{230} - \alpha_U - \lambda_{238}},$$

(5.59)

$$c_5 = \frac{\alpha_U + \lambda_{238} + \theta}{\left(\alpha_{Ra} + \lambda_{226} + \theta\right)\left(\alpha_{Ra} + \lambda_{226} - \alpha_{Th} - \lambda_{230}\right)} \times$$
$$\left(\alpha_{Ra} - \alpha_{Th} - \lambda_{230} + \frac{\lambda_{230}\lambda_{226}}{\alpha_{Ra} + \lambda_{226} - \alpha_U - \lambda_{238}}\right).$$

(5.60)

Therefore, $\left(^{230}Th/^{238}U\right)$ and $\left(^{226}Ra/^{230}Th\right)$ in the extracted melt are given by

$$\frac{\left(^{230}Th\right)}{\left(^{238}U\right)} = \frac{F_{Th}}{F_U}\left\{c_1 + c_2\frac{1 - \exp\left[-\left(\alpha_{Th} + \lambda_{230} + \theta\right)T\right]}{1 - \exp\left[-\left(\alpha_U + \lambda_{238} + \theta\right)T\right]}\right\},$$

(5.61)

$$\frac{\left(^{226}Ra\right)}{\left(^{230}Th\right)} = \frac{F_{Ra}}{F_{Th}} \times \frac{\begin{cases} c_3\left\{1-\exp\left[-(\alpha_U + \lambda_{238} + \theta)T\right]\right\} + \\ c_4\left\{1-\exp\left[-(\alpha_{Th} + \lambda_{230} + \theta)T\right]\right\} + \\ c_5\left\{1-\exp\left[-(\alpha_{Ra} + \lambda_{226} + \theta)T\right]\right\} \end{cases}}{\begin{cases} c_1\left\{1-\exp\left[-(\alpha_U + \lambda_{238} + \theta)T\right]\right\} + \\ c_2\left\{1-\exp\left[-(\alpha_{Th} + \lambda_{230} + \theta)T\right]\right\} \end{cases}}. \tag{5.62}$$

The system of differential equations (5.15) and (5.16) for ^{238}U-^{230}Th systematics (rather than that of Eqs. (5.16) and (5.17) for ^{230}Th -^{226}Ra systematics) is also suitable for ^{235}U-^{231}Pa systematics, therefore,

$$\frac{\left(^{231}Pa\right)}{\left(^{235}U\right)} = \frac{F_{Pa}}{F_U}\left\{c_6 + c_7\frac{1-\exp\left[-(\alpha_{Pa} + \lambda_{231} + \theta)T\right]}{1-\exp\left[-(\alpha_U + \lambda_{235} + \theta)T\right]}\right\}, \tag{5.63}$$

where

$$c_6 = \frac{\lambda_{231}}{\alpha_{Pa} + \lambda_{231} - \alpha_U - \lambda_{235}}, \tag{5.64}$$

$$c_7 = \frac{(\alpha_U + \lambda_{235} + \theta)(\alpha_{Pa} - \alpha_U - \lambda_{235})}{(\alpha_{Pa} + \lambda_{231} + \theta)(\alpha_{Pa} + \lambda_{231} - \alpha_U - \lambda_{235})}. \tag{5.65}$$

λ_{235} and λ_{231} are decay constants of ^{235}U and ^{231}Pa, respectively. Note that c_1 to c_7 are functions of ϕ and $\dot{M}$ only and they are dimensionless because λ, θ, and α have the same dimension. Equations (5.61), (5.62) and (5.63) are from Zou and Zindler (2000) and are the general equations that will be used for modeling here.

Two limit cases (when $T \to 0$ or $T \to \infty$) for $\left(^{230}Th/^{238}U\right)$, $\left(^{226}Ra/^{230}Th\right)$ and $\left(^{231}Pa/^{235}U\right)$ can be obtained from the general equations (5.61), (5.62) and (5.63). When $T \to 0$, Eqs. (5.61), (5.62) and (5.63) become Eqs. (5.51), (5.52) and (5.53), which are the initial conditions, respectively. On the other hand, when $T \to \infty$, Eqs. (5.61), (5.62) and (5.63) are only functions of ϕ and $\dot{M}$ and reduce to:

$$\frac{\left(^{230}Th\right)_{\infty}}{\left(^{238}U\right)_{\infty}}\left(\phi,\dot{M}\right)=\frac{F_{Th}}{F_U}\left(\frac{\alpha_U+\lambda_{238}+\lambda_{230}+\theta}{\alpha_{Th}+\lambda_{230}+\theta}\right),\qquad(5.66)$$

$$\frac{\left(^{226}Ra\right)_{\infty}}{\left(^{230}Th\right)_{\infty}}\left(\phi,\dot{M}\right)=\frac{F_{Ra}}{F_{Th}}\frac{\alpha_{Th}+\lambda_{226}+\theta}{\alpha_{Ra}+\lambda_{226}+\theta}\times$$

$$\left[1+\frac{\lambda_{230}\left(\alpha_U+\lambda_{238}-\alpha_{Th}\right)}{\left(\alpha_{Th}+\lambda_{226}+\theta\right)\left(\alpha_U+\lambda_{238}+\lambda_{230}+\theta\right)}\right],$$

$$(5.67)$$

$$\frac{\left(^{231}Pa\right)_{\infty}}{\left(^{235}U\right)_{\infty}}\left(\phi,\dot{M}\right)=\frac{F_{Pa}}{F_U}\left(\frac{\alpha_U+\lambda_{235}+\lambda_{231}+\theta}{\alpha_{Pa}+\lambda_{231}+\theta}\right).\qquad(5.68)$$

5.4. Forward Modeling Results

The distribution coefficients of the parent and daughter nuclides are important in U-series disequilibria. If the parent has a higher distribution coefficient than the daughter, its extraction from the matrix is retarded and the daughter/parent activity ratio in the melt can be greater than 1. Experiments have shown that $D_U > D_{Th}$ for garnet peridotites but $D_U < D_{Th}$ for spinel peridotites at low pressures. Due to experimental difficulties, D_{Ra} and D_{Pa} are not directly known but are inferred to be very small. Therefore, it is reasonable to assume $D_{Th} > D_{Ra}$ and $D_U > D_{Pa}$ for both garnet peridotites and spinel peridotites.

From three parameters ϕ, $\dot{M}$ and T, we can calculate the fraction of extracted melt (X) using Eq. (5.44) and three daughter/parent activity ratios using (5.61), (5.62) and (5.63), respectively. In this way, we can make activity ratios vs. X plot. At low X, there is greater disequilibrium due to a strong component of net elemental fractionation of parent and daughter at low degrees of melting. The disequilibrium becomes smaller at high X as a result of small net elemental fractionation, and the disequilibrium is dominated by the component of in-growth of daughter nuclides. The form of the curves in Figs. 5.3, 5.4 and 5.5 reflects the transition from the former to the later with increasing X. In addition, small porosity and slow melting rate favor high $\left(^{230}Th/^{238}U\right)$, $\left(^{226}Ra/^{230}Th\right)$ and $\left(^{231}Pa/^{235}U\right)$ in magmas (Fig. 5.3, 5.4, 5.5), which is consistent with previous assessment by McKenzie (1985) about the effect of melting rate and porosity on $\left(^{230}Th/^{238}U\right)$. However, the sensitivity of the daughter/parent activity ratio to the variations of ϕ and $\dot{M}$ is different for the above three activity ratios. Both $\left(^{230}Th/^{238}U\right)$ and $\left(^{231}Pa/^{235}U\right)$ are sensitive to the variations of ϕ and $\dot{M}$ (5.3, 5.5). In contrast, $\left(^{226}Ra/^{230}Th\right)$ is sensitive to ϕ but very insensitive to $\dot{M}$ and, in particular, X (Fig. 5.4).

Figures 5.3, 5.4 and 5.6 show that the relative magnitude of the daughter and parent distribution coefficients actually determines whether a correlation is positive or negative between the daughter/parent activity ratio and any one of the three parameters (ϕ, $\dot{M}$, and X). If the distribution coefficient of the daughter is smaller than that of its parent, a negative correlation is shown between the daughter/parent activity ratio and any of the above three parameters. For example, if $D_{Pa} < D_U$, then short melting time, small porosity, and slow melting rate favor the generation of high $\left(^{231}Pa/^{235}U\right)$ values, and vice versa.

Equation (5.67) for $\left(^{226}Ra/^{230}Th\right)_\infty$ and Eq. (5.68) for $\left(^{231}Pa/^{235}U\right)_\infty$ give the minimum values for $\left(^{226}Ra/^{230}Th\right)$ and $\left(^{231}Pa/^{235}U\right)$, respectively. As for Eq. (5.66) for $\left(^{230}Th/^{238}U\right)_\infty$, it gives the minimum values if melting takes place in a garnet peridotite source ($D_{Th} < D_U$) and a maximum value for melting of a spinel peridotite source ($D_U < D_{Th}$). In addition, Eq. (5.67) for $\left(^{226}Ra/^{230}Th\right)_\infty$ in general gives a value that is very close to the exact solution from Eq. (5.62). Equations (5.66) and (5.68) give approximately correct daughter/parent activity ratios when the melting time is long or when the fraction of the extracted melt is high. According to Figs. 1 and 2, when $X > 4\%$, the values calculated from Eqs. (5.66) and (5.68) would be very close to the exact solution calculated from Eqs. (5.61) and

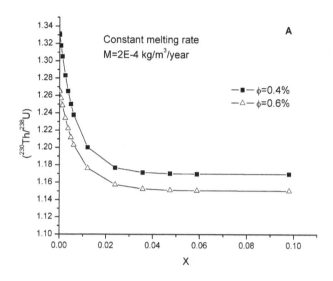

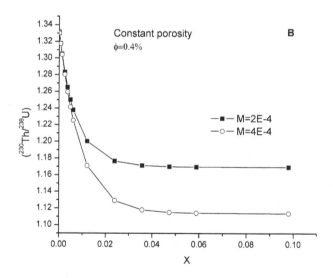

Fig. 5.4. (A) Variations of (^{230}Th/^{238}U) as functions of X and ϕ at a constant $\dot{M}$ (= 2×10^{-4} kg/m^3/year); (B) variation of (^{230}Th/^{238}U) as functions of X and $\dot{M}$ for a constant ϕ (= 0.4%). Source rocks: garnet peridotities. The bulk distribution coefficients are D_{Th} = 0.00353 and D_U = 0.00583 for garnet peridotities.

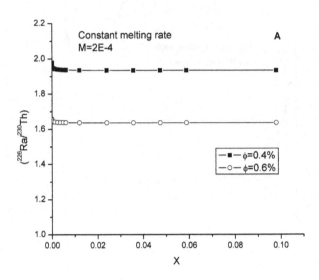

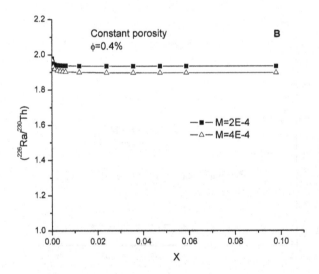

Fig. 5.5. (A) Variations of (^{226}Ra/^{230}Th) as functions of X and ϕ at a constant $\dot{M}$ (= 2×10^{-4} kg/m^3/year); (B) variation of (^{226}Ra/^{230}Th) as functions of X and $\dot{M}$ for a constant ϕ (= 0.4%). The bulk distribution coefficients are D_{Th} = 0.00353 and D_{Ra} = 0.0001.

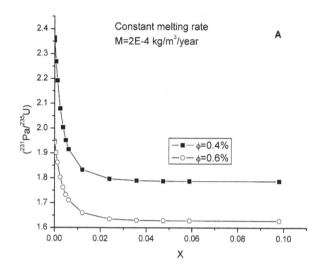

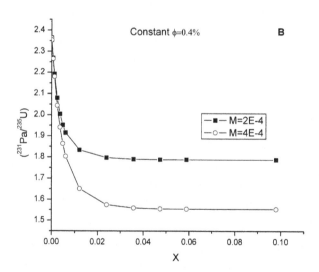

Fig. 5.6. (A) Variations of $(^{231}\text{Pa}/^{235}\text{U})$ as functions of X and ϕ at a constant $\dot{M}$ (= 2×10^{-4} kg/m^3/year); (B) variation of $(^{231}\text{Pa}/^{235}\text{U})$ as functions of X and $\dot{M}$ for a constant ϕ (= 0.4%). The bulk distribution coefficients are D_U = 0.00353 and D_Pa = 0.0005.

(5.63), respectively. The characteristic X_c after which (5.61) and (5.63) return values that are very close to the exact solution depends on the choices of distribution coefficients, the melting rate, and the porosity. Steady state is not necessarily at infinite melting times and can be reached at a characteristic time.

The values of $\left(^{230}Th/^{238}U\right)_\infty$ and $\left(^{226}Ra/^{230}Th\right)_\infty$ have been frequently used to estimate melting porosity and melting rate. This estimate from the simplified equation is applicable when $X > X_c$. In comparison, the pair of $\left(^{231}Pa/^{235}U\right)_\infty$ and $\left(^{226}Ra/^{230}Th\right)_\infty$ have seldom been used but have a potential advantage over the frequently used pair in that $\left(^{231}Pa/^{235}U\right)_\infty$ is empirically less sensitive to the types of the source rocks (spinel peridotites or garnet peridotites) compared with $\left(^{230}Th/^{238}U\right)_\infty$. The disadvantage of using $\left(^{231}Pa/^{235}U\right)_\infty$ instead of $\left(^{230}Th/^{238}U\right)_\infty$ at present lies in the paucity of the data for $\left(^{231}Pa/^{235}U\right)_\infty$ in literature.

5.5. Discussions

5.5.1. The Meanings of melting rate and melt extraction rate

Although the melting rate $\dot{M}$ plays an important role in U-series modeling, this parameter has not been rigorously defined in literature, which may have caused confusion in terms of understanding the melting process. Essentially, in order for Eq. (5.30) to satisfy mass balance requirements for the dynamic melting model, $\dot{M}$ should be defined as

$$\dot{M} = -\frac{1}{1-X}\frac{d\left[\rho_s(1-\phi)(1-X)\right]}{dt} \, , \tag{5.69}$$

where $\rho_s(1-\phi)(1-X)$ is the mass of the residual solid per unit volume. Note that $\dot{M}$ is a scaled parameter relative to the amount of the total residue $(1-X)$ and has the dimension of mass per unit volume per unit time (e.g., kg/m³/year). If we divide $\dot{M}$ by the density of the total residue, or, $\rho_f \phi + \rho_s (1-\phi)$, we can obtain a melting rate in per cent (of the total residue) per unit time as

$$\dot{M}' = -\frac{\rho_s (1-\phi)}{\rho_f \phi + \rho_s (1-\phi)} \frac{1}{1-X} \frac{d(1-X)}{dt}. \qquad (5.70)$$

$\dot{M}'$ has a simpler dimension and is easy to understand as compared to $\dot{M}$.

According to Eq. (5.42), the rate of melt extraction ($\dot{M}_e$) is

$$\dot{M}_e = -\frac{1}{1-X} \frac{d\left\{\left[\rho_f \phi + \rho_s (1-\phi)\right](1-X)\right\}}{dt}, \qquad (5.71)$$

where $\left[\rho_f \phi + \rho_s (1-\phi)\right](1-X)$ is the mass of total residue per unit volume and $\dot{M}_e$ is also a scaled parameter relative to the amount of the total residue. The melt extraction rate in percent (of the total residue) per unit time can be obtained by dividing $\dot{M}_e$ by the density of the total residue, or, $\rho_f \phi + \rho_s (1-\phi)$, we have

$$\dot{M}_e' = -\frac{1}{1-X} \frac{d(1-X)}{dt}. \qquad (5.72)$$

Comparison of Eq. (5.72) with Eq. (5.71) yields the relationship between $\dot{M}$ and $\dot{M}_e$ as

$$\frac{\dot{M}}{\dot{M}_e} = \frac{\rho_s (1-\phi)}{\rho_f \phi + \rho_s (1-\phi)}, \qquad (5.73)$$

which is Eq. 9 in Zou (1998) obtained from a different approach.

5.5.2 U-series modeling and the degree of partial melting

Equations (5.61), (5.62) and (5.63) express the variation of $\left(^{230}Th/^{238}U\right)$, $\left(^{226}Ra/^{230}Th\right)$, and $\left(^{231}Pa/^{235}U\right)$ in the extracted melt as functions of ϕ, $\dot{M}$, and T (ϕ and $\dot{M}$ are implicitly included in the melting parameter α). One might wonder why some commonly used parameters in trace element modeling, such as the degree of partial melting (f) or the fraction of extracted melt (X), are not included in these equations. This is due to the fact that F and X are also functions of ϕ, $\dot{M}$, and T for U-series modeling. The expression for X as a function of ϕ, $\dot{M}$, and T, has been given by Eq. (5.44). And f is related to X by (Zou, 1998)

$$f = \Phi + (1 - \Phi) X, \qquad (5.74)$$

where Φ is the mass porosity and is related to volume porosity (ϕ) by

$$\Phi = \frac{\rho_f \phi}{\rho_f \phi + \rho_s (1 - \phi)}. \qquad (5.75)$$

For U-series modeling, we have five parameters in ϕ, $\dot{M}$, T, f and X. Since there are two equations (Eqs. (5.44) and (5.74)) to relate these five parameters, we have three independent parameters. Theoretically, we can express $\left({}^{230}Th / {}^{238}U \right)$, $\left({}^{226}Ra / {}^{230}Th \right)$ and $\left({}^{231}Pa / {}^{235}U \right)$ as functions of any three of the above five parameters except for the combination of ϕ, f and X. For example, we may select ϕ, $\dot{M}$, and f as independent parameters. By combining Eqs. (5.44) and (5.74), we have

$$f = \Phi + (1 - \Phi) \left\{ 1 - \exp \left[- \frac{\dot{M} T}{\rho_s (1 - \phi)} \right] \right\}, \qquad (5.76)$$

or

$$T = \frac{\rho_s (1 - \phi)}{\dot{M}} \ln \left(\frac{1 - \Phi}{1 - f} \right). \qquad (5.77)$$

Substitution of Eq. (5.77) into Eqs. (5.61), (5.62) and (5.63) lead to the formula for $\left({}^{230}Th / {}^{238}U \right)$, $\left({}^{226}Ra / {}^{230}Th \right)$ and $\left({}^{231}Pa / {}^{235}U \right)$ as functions of ϕ, $\dot{M}$, and f. However, doing so will make these equations unnecessarily complex. In practice, the selection of ϕ, $\dot{M}$, and T leads to the simplest formula.

The physical significance of melting time has not been well handled before. Equation (5.77) is in fact the definition of melting time. It can be seen from Eq. (5.77) that if $f = \Phi$, then $T = 0$; and if $f \to 1$, then $T \to \infty$. Accordingly, increasing melting time reflects increasing degrees of melting in these time-dependent melting calculations, and the scenario of infinite melting time corresponds to total melting. Therefore, melting time is the time period for the degree of partial melting increases from Φ to its final degree of melting at which the measured samples are produced. Physically, given a certain melting rate, the low melting times represent either simply a small amount of melting at low beta factors, or volcanism at the onset of complete rifting.

5.5.3. Activity ratios in the melt when $f < \Phi$ before melt extraction starts

Equations (5.61), (5.62) and (5.63) describe the activity ratios in the extracted melt when $f > \Phi$. In this section we will discuss the behaviors of activity ratios when $f \leq \Phi$. Since there is no extracted melt when $f \leq \Phi$, we only study the behaviors in the residual melt. We will first describe the case when $f = \Phi$ (or the initial conditions) and then the case when $f < \Phi$. According to Eq.(5.33), we have

$$F_{Th}/F_U = \frac{D_U \rho_s (1-\phi) + \rho_f \phi}{D_{Th} \rho_s (1-\phi) + \rho_f \phi},$$ (5.78)

Since the effective distribution coefficients for U and Th are

$$D_U^{eff} = \Phi + (1-\Phi)D_U,$$ (5.79)

$$D_{Th}^{eff} = \Phi + (1-\Phi)D_{Th},$$ (5.80)

where mass porosity Φ is related to volume porosity ϕ by Eq. (5.75). According to equations (5.78), (5.79) and (5.80), we obtain

$$F_{Th}/F_U = D_U^{eff}/D_{Th}^{eff}.$$ (5.81)

Substitution of (5.81) into Eq. (5.51) for the initial activity ratio gives rise to

$$\left(^{230}Th\right)_f^0 \Big/ \left(^{238}U\right)_f^0 = D_U^{eff}/D_{Th}^{eff} = \frac{\Phi + (1-\Phi)D_U}{\Phi + (1-\Phi)D_{Th}}.$$ (5.82)

Consequently, the initial ratio $\left(^{230}Th\right)_f^0 \Big/ \left(^{238}U\right)_f^0$ is actually the inverse ratio of effective distribution coefficient of Th over that of U. According to Eq. (5.82), the initial ratio when $f = \Phi$ is controlled by D_U, D_{Th} and Φ, and is independent of melting rate.

As for the other case when $f < \Phi$, melting takes place in a closed system, and the melting process is the same as batch melting (Fig. 1.2). Replacing Φ in Eq. (5.82) by f, we can obtain the equation for $f < \Phi$,

$$\left(^{230}Th\right)_f \Big/ \left(^{238}U\right)_f = \frac{f + (1-f)D_U}{f + (1-f)D_{Th}}.$$ (5.83)

Therefore, when $f \leq \Phi$, $\left(^{230}Th\right)_f \Big/ \left(^{238}U\right)_f$ in the residual melt only depends on the bulk distribution coefficients of Th and U and the degree of melting (f), and is independent of melting rate, which is unlike the case when $f > \Phi$.

5.5.4. U-Series modeling versus trace element modeling

There is fundamental difference between trace element modeling and U-series modeling of the extracted melt. Melting rate affects U-series systematics but not stable element concentrations. Here we will explore this subject quantitatively. According to Eq. (5.50), trace element concentration in the extracted melt is a function of ϕ and X (or ϕ and f) and is independent of the melting rate ($\dot{M}$). In comparison, on the basis of Eqs. (5.61), (5.62) and (5.63), the activities of U-series nuclides in the extracted melt depend on the rate of melting. Although ^{238}U and ^{235}U can be treated as stable nuclides for the time scale of melting because of their long half lives, we can not treat $\left(^{230}Th\right)$, $\left(^{226}Ra\right)$, and $\left(^{231}Pa\right)$ the same way. This is due to the fact that, unlike ^{238}U and ^{235}U, the decay constants for ^{230}Th, ^{226}Ra and ^{231}Pa are too large to be ignored when compared with the melting parameter α (see, e.g., Eqs. (5.61), (5.62) and (5.63)).

The difference between U-series modeling and trace element modeling can also be shown graphically. We take U-Th disequilibrium as an example. Figure 5.8A compares the relative extent of U/Th fractionation versus ingrowth as a function of melting rate. It can be seen that, for a given fraction of extracted melt and a given porosity,

$\left(^{230}Th/^{238}U\right)$ strongly depends on the melting rate whereas the source-normalized Th/U concentration ratio, $[Th/U]_{melt}/[Th/U]_{source}$, is independent of the melting rate and is significantly lower than $\left(^{230}Th/^{238}U\right)$. Figure 5.8A can also elucidate an important point that could be tied back into existing data sets. For example, Hawaiian basalts have a large buoyancy flux (i.e. upwelling rate) and little growth (Cohen and O'Nions, 1993; Sims et al., 1995) while MORBs have much lower upwelling rates and considerable ^{230}Th ingrowth (e.g., Bourdon et al., 1996).

Both net elemental U-Th fractionation and ^{230}Th ingrowth contribute to the total U-Th disequilibrium. The total U-Th disequilibrium can be expressed as $\left(^{230}Th/^{238}U\right)-1$, where $\left(^{230}Th/^{238}U\right)$ is calculated from Eq. (5.61). Net elemental U-Th fractionation can be expressed as $[Th/U]_{melt}/[Th/U]_{source}-1$, where $[Th/U]_{melt}/[Th/U]_{source}$ can be calculated from Eq. (5.50). Consequently, the ^{230}Th ingrowth can be expressed as:

$$\left(^{230}Th/^{238}U\right) - [Th/U]_{melt}/[Th/U]_{source}.$$

The relative contribution from net elemental U-Th fractionation to the total U-Th disequilibrium can be quantified as

$$\kappa = \frac{[Th/U]_{melt}/[Th/U]_{source}-1}{\left(^{230}Th/^{238}U\right)-1}. \tag{5.84}$$

The relative contribution of ^{230}Th ingrowth to total U-Th disequilibrium is thus

$$1-\kappa = \frac{\left(^{230}Th/^{238}U\right)-[Th/U]_{melt}/[Th/U]_{source}}{\left(^{230}Th/^{238}U\right)-1}. \tag{5.85}$$

The relative contribution from net elemental U-Th fractionation, κ, strongly depends on the melting rate and its magnitude is usually significantly less than 1 (Fig. 5.8B). Net elemental U-Th fractionation is significant only when the melting rate is large.

It should be pointed out that κ is also strongly dependent on the decay constant of the daughter nuclide. The larger the decay constant of the daughter isotope (i.e., the shorter its half life), the smaller the κ. This is one of the reasons why ^{230}Th-^{226}Ra and ^{235}U-^{231}Pa disequilibria

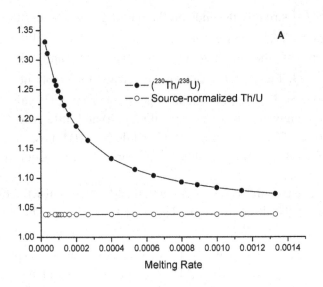

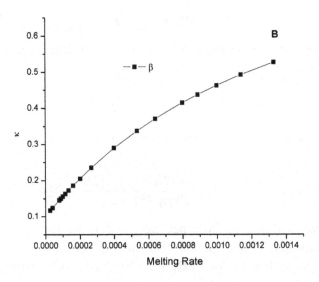

Fig. 5.8. (A) Variation of $(^{230}\text{Th}/^{238}\text{U})$ in the extracted melt as a function of $\dot{M}$ (kg/m³/year) for a fixed ϕ (0.3%) and X (2.4%) produced by melting of a garnet peridotite source. The source-normalized Th/U represents the Th/U ratio in the extracted melt normalized to its source Th/U ratio and is calculated using Eq. (5.50) for stable

elements. (^{230}Th/^{238}U) decreases as the melting rate increases, in contrast, $[Th/U]_{melt}/[Th/U]_{source}$ is independent of the rate of melting and is significantly smaller than (^{230}Th/^{238}U). (B) The relative contribution of net elemental U-Th fractionation to the total ^{230}Th-^{238}U disequilibrium as a function of melting rate for a fixed ϕ (0.3%) and X (2.4%). The bulk distribution coefficients are D_{Th} = 0.00353 and D_U = 0.00583 for garnet peridotites.

are more pronounced than ^{238}U-^{230}Th disequilibrium in young lavas. For example, mid-ocean ridge basalts often have (^{231}Pa/^{235}U) about 2.5 but have (^{230}Th/^{238}U) less than 1.40.

Since κ is usually significantly less than 1, we cannot use the equations for trace element fractionation to model $\left(^{230}Th\right)$, $\left(^{226}Ra\right)$, $\left(^{231}Pa\right)$, $\left(^{230}Th/^{238}U\right)$, $\left(^{226}Ra/^{230}Th\right)$, or $\left(^{231}Pa/^{235}U\right)$ by assuming that net elemental fractionation alone is responsible for the U-series disequilibria. Otherwise, the inferred degree of partial melting to produce basalts based on elemental fractionation alone would be unrealistically too small.

5.6. Chromatographic Melt Transport

The dynamic melting model emphasizes the variation of concentration with time (dc/dt) but ignores the change of concentration with spatial position (dc/dz), where z is the depth in a one-dimensional melting column. In contrast, the steady-state chromatographic melt transport model takes into account dc/dz but assumes $dc/dt = 0$ (steady-state). The two models also differ in the extents of chemical equilibration after melt extraction. The dynamic melting model assumes no re-equilibration of extracted melt with the matrix during melt migration to the surface, whereas chromatographic transport model assumes instant re-equilibrium with matrix throughout the transport process.

According to Spiegelman and Elliott (1993), the consequences of melt transport through a steady-state ($dc_i/dt = 0$), one-dimensional melting column on the concentration of radioacitve isotopes can be separated from batch melting by expressing the concentration of an element as:

$$\alpha_i = \frac{c_i}{c_{bi}}. \tag{5.86}$$

c_i is the melt concentration divided by the initial source concentration, c_{bi} is the concentration of a stable element due to batch melting divided by the initial source concentration and α_i is the enrichment factor due to melt transport over batch melting:

$$c_{bi} = \frac{1}{D + (1-D)F_{\max}\zeta}. \tag{5.87}$$

(note there is a typographical error in Eq. 13 of Spiegelman and Elliott (1993) for c_{bi})

In a one-dimensional steady-state melting column, the variation of the enrichment factor α_i with depth due to melt transport is given by Spiegelman and Elliott (1993) (after correction of another typographical error in their Eq. 15)

$$\frac{d\alpha_i}{d\zeta} = \lambda h \left(\frac{\alpha_{i-1}}{w_{eff}^{i-1}} - \frac{\alpha_i}{w_{eff}^{i}} \right), \tag{5.88}$$

where

$$w_{eff}^{i} = \frac{\rho_f \phi w + \rho_s (1-\phi) D_i W}{\rho_f \phi + \rho_s (1-\phi) D_i}. \tag{5.89}$$

Let

$$\tau = h / w_{eff}, \tag{5.90}$$

we have

$$\frac{d\alpha_i}{d\zeta} = \lambda_i \left(\alpha_{i-1}\tau_{i-1} - \alpha_i \tau_i \right). \tag{5.91}$$

For the decay chain ^{238}U-^{230}Th-^{226}Ra, we have

$$\frac{d\alpha_0}{d\zeta} = -\lambda_0 \alpha_0 \tau_0 , \tag{5.92}$$

$$\frac{d\alpha_1}{d\zeta} = \lambda_1 \left(\alpha_0 \tau_0 - \alpha_1 \tau_1 \right), \tag{5.93}$$

$$\frac{d\alpha_2}{d\zeta} = \lambda_2 \left(\alpha_1 \tau_1 - \alpha_2 \tau_2 \right), \tag{5.94}$$

where subscripts 0, 1 and 2 denote ^{238}U, ^{230}Th, ^{226}Ra , respectively. The analytical solutions to Eqs. (5.92), (5.93) and (5.94) are given by (Sims et al., 1999)

$$\alpha_0 = \alpha_0 e^{-\lambda_0 \tau_0 \zeta} , \tag{5.95}$$

$$\alpha_1(\zeta) = \frac{\lambda_1 \tau_0 \alpha_0^0}{\lambda_1 \tau_1 - \lambda_0 \tau_0} e^{-\lambda_0 \tau_0 \zeta} + \left(\alpha_1^0 - \frac{\lambda_1 \tau_0 \alpha_0^0}{\lambda_1 \tau_1 - \lambda_0 \tau_0} \right) e^{-\lambda_1 \tau_1 \zeta} , \tag{5.96}$$

$$\alpha_2(\zeta) = \frac{\lambda_1 \lambda_2 \tau_1 \tau_0 \alpha_0^0}{(\lambda_1 \tau_1 - \lambda_0 \tau_0)(\lambda_2 \tau_2 - \lambda_0 \tau_0)} e^{-\lambda_0 \tau_0 \zeta}$$

$$+ \frac{\lambda_2 \tau_1}{\lambda_2 \tau_2 - \lambda_1 \tau_1} \left(\alpha_1^0 - \frac{\lambda_1 \tau_0 \alpha_0^0}{\lambda_1 \tau_1 - \lambda_0 \tau_0} \right) e^{-\lambda_1 \tau_1 \zeta} \tag{5.97}$$

$$+ \left[\begin{array}{c} \alpha_2^0 - \dfrac{\lambda_1 \lambda_2 \tau_1 \tau_0 \alpha_0^0}{(\lambda_1 \tau_1 - \lambda_0 \tau_0)(\lambda_2 \tau_2 - \lambda_0 \tau_0)} \\ - \dfrac{\lambda_2 \tau_1}{\lambda_2 \tau_2 - \lambda_1 \tau_1} \left(\alpha_1^0 - \dfrac{\lambda_1 \tau_0 \alpha_0^0}{\lambda_1 \tau_1 - \lambda_0 \tau_0} \right) \end{array} \right] e^{-\lambda_2 \tau_2 \zeta} .$$

Equations (5.95), (5.96) and (5.97) are suitable for constant critical melting porosity. In a one dimensional steady state melting column as a result of decompression melting, the porosity may increase from the bottom to the top of the column. If melting porosities change as a function of the spatial position, the related differential equations need to be solved numerically. More details of various melt transport models by porous flow have been given by Spiegelman and Elliott (1993), Iwamori, (1994), and Lundstrom (2000).

5.7. Summary of Equations

1) In-situ decay, no initial daughter nuclides, no melting (Bateman Equation)

$$^{238}U = {}^{238}U_0 e^{-\lambda_{238}t},$$

$$^{230}Th = \frac{\lambda_{238}}{\lambda_{230} - \lambda_{238}} {}^{238}U_0 \left(e^{-\lambda_{238}t} - e^{-\lambda_{230}t}\right),$$

$$^{226}Ra = \frac{\lambda_{230}\lambda_{238}}{\lambda_{230} - \lambda_{238}} {}^{238}U_0 \left[\frac{e^{-\lambda_{238}t} - e^{-\lambda_{226}t}}{\lambda_{226} - \lambda_{238}} - \frac{e^{-\lambda_{230}t} - e^{-\lambda_{226}t}}{\lambda_{226} - \lambda_{230}}\right].$$

2) In-situ decay, with initial daughter nuclides, no melting

$$^{238}U = {}^{238}U_0 e^{-\lambda_{238}t},$$

$$^{230}Th = \frac{\lambda_{238}}{\lambda_{230} - \lambda_{238}} {}^{238}U_0 \left(e^{-\lambda_{238}t} - e^{-\lambda_{230}t}\right) + {}^{230}Th_0 e^{-\lambda_{230}t},$$

$$^{226}Ra = {}^{226}Ra_0 \, e^{-\lambda_{226}t}$$

$$+ \frac{\lambda_{230}}{\lambda_{226} - \lambda_{230}} {}^{230}Th_0 \left[e^{-\lambda_{230}t} - e^{-\lambda_{226}t}\right]$$

$$+ \frac{\lambda_{230}\lambda_{238}}{\lambda_{230} - \lambda_{238}} {}^{238}U_0 \left[\frac{e^{-\lambda_{238}t} - e^{-\lambda_{226}t}}{\lambda_{226} - \lambda_{238}} - \frac{e^{-\lambda_{230}t} - e^{-\lambda_{226}t}}{\lambda_{226} - \lambda_{230}}\right].$$

3) Dynamic melting with initial secular equilibrium

For the instantaneous melt, the activities are given by (5.37), (5.38) and (5.39). For the extracted melt, the activity ratios are provided by (5.61), (5.62) and (5.63). At high degree of melting, the equations for activity ratios can be simplified as:

$$\frac{\left(^{230}Th\right)_{\infty}}{\left(^{238}U\right)_{\infty}}\left(\phi,\dot{M}\right) = \frac{F_{Th}}{F_{U}}\left(\frac{\alpha_{U}+\lambda_{238}+\lambda_{230}+\theta}{\alpha_{Th}+\lambda_{230}+\theta}\right),$$

$$\frac{\left(^{226}Ra\right)_{\infty}}{\left(^{230}Th\right)_{\infty}}\left(\phi,\dot{M}\right) = \frac{F_{Ra}}{F_{Th}}\frac{\alpha_{Th}+\lambda_{226}+\theta}{\alpha_{Ra}+\lambda_{226}+\theta}\times$$

$$\left[1+\frac{\lambda_{230}\left(\alpha_{U}+\lambda_{238}-\alpha_{Th}\right)}{\left(\alpha_{Th}+\lambda_{226}+\theta\right)\left(\alpha_{U}+\lambda_{238}+\lambda_{230}+\theta\right)}\right],$$

$$\frac{\left(^{231}Pa\right)_{\infty}}{\left(^{235}U\right)_{\infty}}\left(\phi,\dot{M}\right) = \frac{F_{Pa}}{F_{U}}\left(\frac{\alpha_{U}+\lambda_{235}+\lambda_{231}+\theta}{\alpha_{Pa}+\lambda_{231}+\theta}\right).$$

4) Melt transport with initial secular equilibrium

For melt as porous flow, the enrichment factors due to melt transport over batch melting at variable depths in a one dimensional melting column are given by (5.95), (5.96) and (5.97).

References

Bourdon, B., Zindler, A., Elliott, T. and Langmuir, C.H., 1996. Constraints on mantle melting at midocean ridges from global ^{238}U-^{230}Th disequilibrium data. Nature, 384: 231-235.

Chabaux, F. and J., A.C., 1994. ^{238}U-^{230}Th-^{226}Ra disequilibria in volcanics: A new insight into melting conditions. Earth Planet. Sci. Lett., 126: 61-74.

Cohen, A.S. and K., O.N.R., 1993. Melting rates beneath Hawaii: Evidence from uranium series isotopes in recent lavas. Earth Planet. Sci. Lett., 120: 169-175.

Iwamori, H., 1994. ^{238}U-^{230}Th-^{226}Ra and ^{235}U-^{231}Pa disequilibria produced by mantle melting with porous and channel flows. Earth Planet. Sci. Lett., 125: 1-16.

Lundstrom, C., 2000. Models of U-series disequilibria generation in MORB: the effects of two scales of melt porosity. Physics of the Earth and Planetary Interiors, 121: 189-204.

McKenzie, D., 1985. ^{230}Th- ^{238}U disequilibrium and the melting processes beneath ridge axes. Earth Planet. Sci. Lett., 72: 149-157.

O'Nions, R.K. and McKenzie, D., 1993. Estimates of mantle thorium/uranium ratios from Th, U and Pb isotope abundances in basaltic melts. Phil. Trans. R. Soc. Lond. A., 342: 65-77.

Sims, K.W.W., DePaolo, D.J., Murrell, M.T., Baldredge, W.S., Goldstein, S., Clague, D. and Jull, M., 1999. Porosity of the melting zone and variations in the solid mantle upwelling rate beneath Hawaii: Inferences from ^{238}U-^{230}Th-^{226}Ra and ^{235}U-^{231}Pa disequilibria. Geochim. Cosmochim. Acta, 63: 4119-4138.

Sims, K.W.W., DePaolo, D.J., Murrell, M.T., Baldridge, W.S., Goldstein, S.J. and Clague, D.A., 1995. Mechanisms of magma generation beneath Hawaii and mid-ocean ridges: Uranium/thorium and samarium/neodymium isotopic evidence. Science, 267: 508-512.

Spiegelman, M. and Elliott, T., 1993. Consequences of melt transport for uranium series disequilibrium in young lavas. Earth Planet. Sci. Lett., 118: 1-20.

Williams, R.W. and Gill, J.B., 1989. Effect of partial melting on the uranium decay series. Geochim. Cosmochim. Acta, 53: 1607-1619.

Zou, H.B., 1998. Trace element fractionation during modal and nonmodal dynamic melting and open-system melting: A mathematical treatment. Geochim. Cosmochim. Acta, 62: 1937-1945.

Zou, H.B. and Zindler, A., 2000. Theoretical studies of ^{238}U-^{230}Th-^{226}Ra and ^{231}Pa-^{235}U disequilibria in young lavas produced during mantle melting. Geochim. Cosmochim. Acta, 64: 1809-1817.

Chapter 6

Crystallization, Assimilation, Mixing

6.1. Crystallization

Although crystallization can be roughly regarded as an inverse process of partial melting, there is some difference. The solid source for partial melting is often a mineralogically heterogeneous whereas the initial magma for crystallization is much more homogeneous. For the process of crystallization, define F as the fraction of initial magma left after fractional crystallization. Note that the meaning of F for crystallization is very different from that for partial melting. The conservation of the total mass and the mass for an element can be summarized in Table 6.1.

Table 6.1. Mass balance for crystallization.

	Initial magma	Remaining magma	Solid
Conservation of total mass	M_0	FM_0	$(1-F)M_0$
Conservation of mass for the element	$C_0 M_0$	$C_L M_0 F$	$\overline{C_s}(1-F)M_0$

The balance of the total mass gives

$$M_L + M_s = M_0 , \qquad (6.1)$$

where M_L is the mass of the remaining magma, M_s the mass of the solid crystallized, and M_0 is the mass of the initial magma.

$$M_L = FM_0,$$ (6.2)

$$M_s = (1 - F)M_0.$$ (6.3)

The mass conservation for a trace element requires

$$m_L + m_s = m_0,$$ (6.4)

where m_L, m_s and m_0 are the mass of the remaining magma, the solid crystallized, and the initial magma, respectively. Thus we have

$$m_L = C_L M_L,$$ (6.5)

$$m_s = \overline{C}_s M_s,$$ (6.6)

$$m_0 = C_0 M_0,$$ (6.7)

where C_L, $\overline{C}_s$, and C_0 are the concentrations of a trace element in the remaining magma, the solid crystallized, and the initial magma. Combination of Eqs. (6.2) to (6.7) results in

$$C_L F + \overline{C}_s (1 - F) = C_0.$$ (6.8)

Equation (6.8) is a fundamental equation for crystallization. D can remain constant or change with F. This fundamental equation for fractional melting is a differential equation whereas that for batch melting is an algebraic equation. If D changes, it is more difficult to derive the equations for fractional melting than batch melting owing to the necessity of solving differential equations for fractional melting.

6.1.1. Fractional crystallization

In many cases, because the crystallizing minerals often have different densities from the magma, the crystals can separate from the magma, which is the scenario for fractional crystallization.

Differentiation of Eq. (6.8) gives

$$d\left(C_L F\right) + d\left[\overline{C_s}(1-F)\right] = 0. \tag{6.9}$$

The magma is in equilibrium with the instantaneous crystal, or, the last crystal of the formed solid, by the following relationship

$$C_s^i = C_L D. \tag{6.10}$$

Note that the remaining magma is in equilibrium with instantaneous crystal, rather than the accumulated crystals that have separated from the magma. The concentration in the instantaneous crystal is related to the concentration in the accumulated fractionated crystals by

$$C_s^i = \frac{d\left[\overline{C_s}(1-F)\right]}{d(1-F)}. \tag{6.11}$$

Combination of (6.10) and (6.11) yields

$$d\left[\overline{C_s}(1-F)\right] = DC_L d(1-F). \tag{6.12}$$

Combination of Eq. (6.9) with Eq. (6.12) gives

$$d\left(C_L F\right) + DC_L d(1-F) = 0. \tag{6.13}$$

Using the chain rule, we have

$$FdC_L + C_L dF - DC_L dF = 0. \tag{6.14}$$

Rewriting Eq. (6.14), we obtain

$$\frac{dC}{C_L} = (D-1)\frac{dF}{F_L}. \tag{6.15}$$

Note that, in the course of the derivation of Eq. (6.15), D is necessary to be a constant.

6.1.1.1. Constant D

Assuming constant D, the solution to Eq. (6.15) with initial magma condition C_0 is

$$C_L = C_0 F^{D-1}. \tag{6.16}$$

This is the well-known Rayleigh fractionation law; it is applicable to geological systems only when the bulk partition coefficients remain constant throughout fractional crystallization.

6.1.1.2. Linear D

If the bulk partition coefficient linearly varies with the percent of the remaining magma (F):

$$D = D_0 + bF, \tag{6.17}$$

then Eq. (6.15) becomes (Greenland, 1970)

$$\frac{dC}{C_L} = (D_0 + bF - 1)\frac{dF}{F_L}. \tag{6.18}$$

The solution to Eq. (6.18) with initial condition $C_L^0 = C_0$ is (Greenland, 1970)

$$C_L = C_0 \exp\left[(D_0 - 1)\mathrm{Ln}F + b(F - 1)\right]. \tag{6.19}$$

Note that when $b = 0$, Eq. (6.19) for linear D collapses to Eq. (6.16) for constant D.

6.1.2. Equilibrium crystallization

When the crystallized crystals always stay with the magma and reach chemical equilibrium, the compositions of the crystals are related to the composition of the melt by the bulk partition coefficient,

$$\overline{C_s} = DC_L. \tag{6.20}$$

Substitution of Eq. (6.20) into Eq. (6.8) gives

$$C_L F + DC_L(1 - F) = C_0. \tag{6.21}$$

The concentration in the remaining magma is

$$C_L = \frac{C_0}{F + D(1 - F)}. \tag{6.22}$$

6.2. Assimilation-fractional Crystallization

During fractional crystallization, the country rocks may assimilate the crystallizing magmas (DePaolo, 1981; O'Hara, 1998). This process is called assimilation-fractional crystallization (AFC) (DePaolo, 1981). The conservation of mass for a trace element requires

$$dm_L + dm_s + dm_a = 0, \tag{6.23}$$

where m_L, m_s, and m_a are the masses of the trace element in the remaining magma, in the solid, and in the assimilating country rock, respectively. Since

$$m_L = C_L M_L, \tag{6.24}$$

$$m_s = \overline{C_s} M_s, \tag{6.25}$$

$$m_a = C_a M_a, \tag{6.26}$$

where C_L, $\overline{C_s}$ and C_a are the concentrations in the magma, in the solid, and in the assimilating rock, respectively, and M_L, M_s and M_a are the total masses of the magma, in the solid, and in the assimilating rock, respectively. Substitution of (6.24), (6.25) and (6.26) into Eq. (6.23) results in

$$d(C_L M_L) + d(\overline{C_s} M_s) + d(C_a M_a) = 0. \tag{6.27}$$

The concentration of the latest formed crystal $d(\overline{C_s} M_s)/dM_s$ is in chemical equilibrium with the magma composition (C_L)

$$C_L = \frac{1}{D} \frac{d(\overline{C_s} M_s)}{dM_s}, \tag{6.28}$$

or

$$d(\overline{C_s} M_s) = D C_L dM_s. \tag{6.29}$$

Substitution of Eq. (6.29) into Eq. (6.27) gives rise to

$$M_L dC_L + C_L dM_L + D C_L dM_s + d(C_a M_a) = 0. \tag{6.30}$$

Assuming constant C_a, we have

$$d(C_a M_a) = C_a d(M_a), \qquad (6.31)$$

and Eq. (6.30) reduces to

$$M_L dC_L + C_L dM_L + DC_L dM_s + C_a dM_a = 0. \qquad (6.32)$$

Since

$$M_L = M_0 F, \qquad (6.33)$$

we have

$$dM_L = M_0 dF. \qquad (6.34)$$

Define

$$r = (M_a^0 - M_a)/M_s, \qquad (6.35)$$

and substituting Eqs. (6.33) and (6.35) into the following mass balance equation:

$$M_0 + M_a^0 = M_s + M_L + M_a, \qquad (6.36)$$

we obtain

$$M_s = \frac{M_0(1 - F)}{1 - r}, \qquad (6.37)$$

$$M_a = M_a^0 - \frac{r}{1 - r} M_0(1 - F). \qquad (6.38)$$

Differentiation of Eq. (6.37) gives

$$dM_s = \frac{-M_0}{1 - r} dF. \qquad (6.39)$$

Differentiation of Eq. (6.38) yields

$$dM_a = \frac{rM_0}{1 - r} dF. \qquad (6.40)$$

Substitution of Eqs. (6.33), (6.34), (6.39) and (6.40) into Eq. (6.32) finally results in

$$FM_0 dC_L + C_L M_0 dF - DC_L \frac{M_0}{1 - r} dF + C_a \frac{rM_0}{1 - r} dF = 0, \qquad (6.41)$$

or

$$FdC_L + \left[\frac{1-r-D}{1-r}C_L + \frac{r}{1-r}C_a \right] dF = 0 . \tag{6.42}$$

The solution to Eq. (6.42) is

$$C_L = C_0 F^{\frac{D+r-1}{1-r}} + \frac{C_a r}{D+r-1}\left(1 - F^{\frac{D+r-1}{1-r}} \right), \tag{6.43}$$

which is equivalent of Eq. 6a in DePaolo (1981). Note that when $r = 0$, Eq. (6.43) collapses to the well-known Rayleigh fractionation law Eq. (6.16). A more complex model to take account of not only mass balance but also energy balance during the AFC processes is documented in Spera and Bohrson (2001).

6.3. Two-component Mixing

6.3.1. Derivation

Here we try to derive the equation to link two measured ratios (u/a and v/b) in a mixture of two components (1 and 2). u/a and v/b can be any concentration ratios, e.g., La/Sm, U/Th, or isotope ratios, e.g., $^{87}Sr/^{86}Sr$, $^{206}Pb/^{204}Pb$. The ratio u/a in a two-component mixture is given by

$$y_m = \left(\frac{u}{a} \right)_m = \frac{u_1 f_1 + u_2(1 - f_1)}{a_1 f_1 + a_2(1 - f_1)}, \tag{6.44}$$

where f_1 is the mass proportion of component one and $(1 - f_1)$ is thus the proportion of component 2 in the mixture.
For component 1, we have

$$y_1 = \frac{u_1}{a_1}, \tag{6.45}$$

and for component 2, we have

$$y_2 = \frac{u_2}{a_2}. \tag{6.46}$$

Substitution of (6.45) and (6.46) into Eq. (6.44) to eliminate numerators u_1 and u_2 gives rise to

$$y_m = \frac{a_1 y_1 f_1 + a_2 y_2 (1 - f_1)}{a_1 f_1 + a_2 (1 - f_1)}. \tag{6.47}$$

Rewriting Eq. (6.47), we have

$$f_1 = \frac{-a_2 y_m + a_2 y_2}{(a_1 - a_2) y_m - a_1 y_1 + a_2 y_2}. \tag{6.48}$$

Similarly, a different ratio (v / b) in the same two-component mixture is given by

$$x_m = \left(\frac{v}{b}\right)_m = \frac{v_1 f_1 + v_2 (1 - f_1)}{b_1 f_1 + b_2 (1 - f_1)}. \tag{6.49}$$

The v / b ratio in component 1 is

$$x_1 = \frac{v_1}{b_1}, \tag{6.50}$$

and v / b ratio in component 2 is

$$x_2 = \frac{v_2}{b_2}. \tag{6.51}$$

Substitution of (6.50) and (6.51) into Eq. (6.49) to eliminate numerators v_1 and v_2 leads to

$$x_m = \frac{b_1 x_1 f_1 + b_2 x_2 (1 - f_1)}{b_1 f_1 + b_2 (1 - f_1)}. \tag{6.52}$$

Equation (6.52) can be re-expressed as

$$f_1 = \frac{-b_2 x_m + b_2 x_2}{(b_1 - b_2) x_m - b_1 x_1 + b_2 x_2}. \tag{6.53}$$

Combination of (6.48) with Eq. (6.53) yields

$$\frac{-a_2 y_m + a_2 y_2}{(a_1 - a_2) y_m - a_1 y_1 + a_2 y_2} = \frac{-b_2 x_m + b_2 x_2}{(b_1 - b_2) x_m - b_1 x_1 + b_2 x_2}. \tag{6.54}$$

Rewriting Eq. (6.54), we find

$$\begin{aligned}(a_2 b_1 y_2 - a_1 b_2 y_1) x_m + (a_1 b_2 - a_2 b_1) x_m y_m \\ + (a_2 b_1 x_1 - a_1 b_2 x_2) y_m + (a_1 b_2 x_2 y_1 - a_2 b_1 x_1 y_2) = 0,\end{aligned} \tag{6.55}$$

which is a hyperbola with

$$A x_m + B x_m y_m + C y_m + D = 0, \tag{6.56}$$

where

$$A = a_2 b_1 y_2 - a_1 b_2 y_1, \tag{6.57}$$

$$B = a_1 b_2 - a_2 b_1, \tag{6.58}$$

$$C = a_2 b_1 x_1 - a_1 b_2 x_2, \tag{6.59}$$

$$D = a_1 b_2 x_2 y_1 - a_2 b_1 x_1 y_2. \tag{6.60}$$

Equation (6.55) is identical to Eq. 11 in Vollmer (1976) and Eq. 2 in Langmuir et al. (1978). The mixing curve resulting from this equation is a hyperbola whose curvature is controlled by the B coefficient.

6.3.2. Analysis of the two-component mixing equation

It is clear from Eq. (6.56) that the value of B controls the behaviors of the equation (Langmuir et al., 1978).
When $B = a_1 b_2 - a_2 b_1 = 0$, Eq. (6.56) reduces to

$$A x_m + C y_m + D = 0,$$

which is a straight line in the y_m vs. x_m plot. When $B \neq 0$, Eq. (6.56) is a hyperbola.
1) For ratio vs. ratio plot, it is a hyperbola unless

$$\frac{a_1}{a_2} = \frac{b_1}{b_2}. \tag{6.61}$$

Figure 6.1 gives an example of a mixing hyperbola in a ratio vs. ratio plot.
Define

$$r = \frac{a_1 b_2}{a_2 b} = \frac{a_1 / b_1}{a_2 / b_2},$$ (6.62)

and the numerical value of r is a function of the extent of the curvature between two points and of the overall curvature of the hyperbolic curve (Langmuir et al., 1978). If $r = 1$, then the mixing curve is a straight line. If $r \neq 1$, the sense of the hyperbolic curvature (concave up or concave down) depends on the relative concentrations of a and b in the end members. As r progressively greater than or less than 1, the hyperbolic form of the function becomes more pronounced (Langmuir et al., 1978). Figure 6.2 shows the variation of the extent of the curvature as a function of r.

2) For ratio vs. concentration plot, $b_1 = b_2 = 1$ and $B = a_1 - a_2$, it is also a hyperbola unless

$$a_1 = a_2.$$

3) For element vs. element plot, $a_1 = a_2 = b_1 = b_2 = 1$, and thus $B = 0$, it is a straight line.

The principle of two-component mixing can be applied to isotope geochemistry.

For ^{143}Nd/^{144}Nd vs. ^{87}Sr/^{86}Sr plot, we have

$$a_1 = (^{144}\text{Nd})_1, \ a_2 = (^{144}\text{Nd})_2,$$

$$b_1 = (^{86}\text{Sr})_1, \ b_2 = (^{86}\text{Sr})_2.$$

Since normally

$$\frac{(^{144}\text{Nd})_1}{(^{144}\text{Nd})_2} \neq \frac{(^{86}\text{Sr})_1}{(^{86}\text{Sr})_2},$$

or

$$\frac{(^{144}\text{Nd})_1}{(^{86}\text{Sr})_1} \neq \frac{(^{144}\text{Nd})_2}{(^{86}\text{Sr})_2},$$

thus two component-mixing does not yield a straight line in the ^{143}Nd/^{144}Nd vs. ^{87}Sr/^{86}Sr plot unless the two components have the same Nd/Sr ratios.

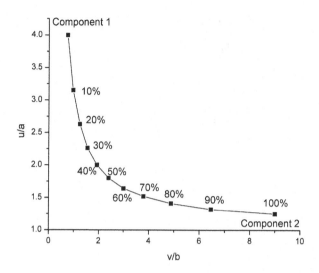

Fig. 6.1. A two-component mixing curve.

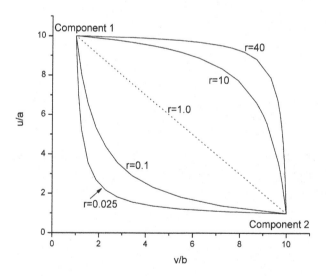

Fig. 6.2. Two-component mixing curves with different r.

For ^{143}Nd/^{144}Nd vs. ^{206}Pb/^{204}Pb plot, we have

$$a_1 = (^{144}\text{Nd})_1, \quad a_2 = (^{144}\text{Nd})_2,$$

$$b_1 = (^{204}\text{Pb})_1, \quad b_2 = (^{204}\text{Pb})_2.$$

Since normally

$$\frac{(^{144}\text{Nd})_1}{(^{144}\text{Nd})_2} \neq \frac{(^{204}\text{Pb})_1}{(^{204}\text{Pb})_2},$$

or

$$\frac{(^{144}\text{Nd})_1}{(^{204}\text{Pb})_1} \neq \frac{(^{144}\text{Nd})_2}{(^{204}\text{Pb})_2},$$

thus, the two-component mixing does not generate a straight line in the ^{143}Nd/^{144}Nd vs. ^{206}Pb/^{204}Pb plot.

For ^{206}Pb/^{204}Pb vs. ^{207}Pb/^{204}Pb plot

$$a_1 = (^{204}\text{Pb})_1, \quad a_2 = (^{204}\text{Pb})_2,$$

$$b_1 = (^{204}\text{Pb})_1, \quad b_2 = (^{204}\text{Pb})_2.$$

Since

$$\frac{a_1}{a_2} = \frac{b_1}{b_2} = \frac{(^{204}\text{Pb})_1}{(^{204}\text{Pb})_2},$$

the two-component mixing always generates a straight line in the ^{206}Pb/^{204}Pb vs. ^{207}Pb/^{204}Pb plot. This is also true for ^{206}Pb/^{204}Pb vs. ^{208}Pb/^{204}Pb plot.

We already know that ^{87}Sr/^{86}Sr vs. Sr is normally not straight line. But ^{87}Sr/^{86}Sr vs. 1/Sr plot is a straight line for two component mixing. For the ^{87}Sr/^{86}Sr vs. 1/Sr plot, we have

$$a_1 = (^{86}\text{Sr})_1, \quad a_2 = (^{86}Sr)_2,$$

$$b_1 = (Sr)_1, \quad b_2 = (Sr)_2.$$

Thus

$$\frac{a_1}{a_2} = \frac{(^{86}\text{Sr})_1}{(^{86}\text{Sr})_2},$$

$$\frac{b_1}{b_2} = \frac{(\text{Sr})_1}{(\text{Sr})_2}.$$

Since

$$\frac{(^{86}\text{Sr})_1}{(^{86}\text{Sr})_2} = \frac{(\text{Sr})_1}{(\text{Sr})_2},$$

we have

$$\frac{(^{86}\text{Sr})_1}{(\text{Sr})_1} = \frac{(^{86}\text{Sr})_2}{(\text{Sr})_2}.$$

Therefore, in the $^{87}\text{Sr}/^{86}\text{Sr}$ vs. $1/\text{Sr}$, the plots fall in a straight line.

6.4. Summary of Equations

1) Crystallization

Fractional crystallization
Constant D:

$$C_L = C_0 F^{D-1}.$$

Linear D:

$$C_L = C_0 \exp\left[(D_0 - 1)\text{Ln}F + b(F - 1)\right].$$

Equilibrium crystallization

$$C_L = \frac{C_0}{F + D(1 - F)}.$$

2) Assimilation-factional crystallization

$$C_L = C_0 F^{\frac{D+r-1}{1-r}} + \frac{C_a r}{D+r-1}\left(1 - F^{\frac{D+r-1}{1-r}}\right).$$

3) Two component mixing

$$Ax_m + Bx_m y_m + Cy_m + D = 0,$$

where

$$A = a_2 b_1 y_2 - a_1 b_2 y_1,$$

$$B = a_1 b_2 - a_2 b_1,$$

$$C = a_2 b_1 x_1 - a_1 b_2 x_2,$$

$$D = a_1 b_2 x_2 y_1 - a_2 b_1 x_1 y_2.$$

The mixing curve is a hyperbola for ratio vs. ratio plot, unless $B = 0$, that is, $a_1 / a_2 = b_1 / b_2$.

References

DePaolo, D.J., 1981. Trace element and isotopic effects of combined wallrock assimilation and fractional crystallization. Earth Planet. SCi. Lett., 1981: 189-202.

Greenland, L.P., 1970. An equation for trace element distribution during magmatic crystallization. Am. Mineral., 55: 455-465.

Langmuir, C.H., Vocke, R.D., Jr. and Hanson, G.N., 1978. A general mixing equition with applications to Icelandic basalts. Earth Planet. Sci. Lett., 37: 380-392.

O'Hara, M.J., 1998. Volcanic plumbing and the space problem-Thermal and geochemical consequences of large-scale assimilation in ocean island development. J. Petrol., 39: 1077-1089.

Spera, F.J. and Bohrson, W.A., 2001. Energy-constrained open-system magmatic processes I: general model and energy-constrained assimilation and fractional crystallization (EC-AFC) formulation. J. Petrol., 42: 999-1018.

Vollmer, R., 1976. Rb-Sr and U-Th-Pb systematics of alkaline rocks: the alkaline rocks from Italy. Geochim. Cosmochim. Acta, 40: 283-295.

Chapter 7

Inverse Geochemical Modeling

Chapters 1-6 present forward modeling of the behaviors of trace elements and isotopes during geological processes. In this chapter, we focus on inverse geochemical modeling. For forward modeling, we may build complex models by adding more parameters into simples ones. We may evaluate the effects of these additional parameters by comparing the complex models with simple ones. For inverse modeling, we try to use relatively simple models to reduce the number of parameters for inversion. Inverse modeling is useful in geochemistry because geochemical investigation often involves deciphering information from the measured elemental abundances and isotope compositions (Allegre and Minster, 1978; Frey et al., 1978; Minster and Allegre, 1978; Hofmann and Feigenson, 1983; McKenzie and O'Nions, 1991; Maaloe, 1994; Zou and Zindler, 1996; Class and Goldstein, 1997).

7.1. Batch Melting Inversion of Melt Compositions

Problem. Demonstrate low-degree (F_1)/high-degree (F_2) concentration ratio (Q) of the same element decreases with increasing partition coefficients (D_0) during modal batch melting.

During batch melting, the concentration in the melt (C_0) is given by

$$C_L = \frac{C_0}{D_0 + F(1 - D_0)} = \frac{C_0}{F + D_0(1 - F)}, \qquad (7.1)$$

where C_0 is the source composition, F is the degree of partial melting, D_0 is the initial bulk partition coefficient.

For batch melting, the low-degree/high degree concentration ratio is given by

$$Q = \frac{C_1}{C_2} = \frac{F_2 + D_0(1 - F_2)}{F_1 + D_0(1 - F_1)}, \qquad (7.2)$$

where $F_1 < F_2$.

By differentiating Q with respect to D_0, we find the first derivative

$$\frac{\partial Q}{\partial D_0} = \frac{(F_1 - F_2)}{\left[F_1 + (1 - F_1) D_0 \right]^2}. \qquad (7.3)$$

Since $F_1 < F_2$, we have $\partial Q / \partial D < 0$. Therefore, Q decreases with increasing D.

7.1.1. Concentration ratio method

Assuming eutectic batch melting, the variation in concentration of a trace element of the melt, C_L, is given by (Shaw, 1970):

$$C_L = \frac{C_0}{D_0 + F(1 - P)}, \qquad (7.4)$$

where C_0 is the concentration of the source, D the distribution coefficient, F the partial melting degree, and P a constant determined by the mineral proportions in the melt.

For element a (highly incompatible) and b (not-so-highly incompatible), their concentrations in the melt C_a and C_b are given by:

$$C_a = \frac{C_a^0}{D_a + F(1 - P_a)}, \qquad (7.5)$$

$$C_b = \frac{C_b^0}{D_b + F(1 - P_b)}. \qquad (7.6)$$

Let the degree of partial melting increase from stage 1 to stage 2. Using Eq. (7.5) for two different degrees of partial melting, F_1 and F_2, the low- F /high- F enrichment ratio Q_a of the highly incompatible elements is given by:

$$Q_a = \frac{C_a^1}{C_a^2} = \frac{D_a + F_2(1-P_a)}{D_a + F_1(1-P_a)}.$$ (7.7)

Similarly, using Eq. (7.6) the enrichment ratio Q_b for the less incompatible element is defined by:

$$Q_b = \frac{C_b^1}{C_b^2} = \frac{D_b + F_2(1-P_b)}{D_b + F_1(1-P_b)}.$$ (7.8)

The above equations (7.7) to (7.8) are all from Maaloe (1994). F_1 and F_2 can be obtained from the system of Eqs (7.7) to (7.8) by making f-Q_b diagrams (see, p. 2520, Maaløe, 1994). Since the F - Q_b diagram is very specific (e.g., each diagram only applies for each Q_a value) and the system is analytical, we prefer explicit solutions and provide them as follows (Zou and Zindler, 1996):

$$F_1 = \frac{D_a(1-P_b)(1-Q_a) - D_b(1-P_a)(1-Q_b)}{(Q_a - Q_b)(1-P_a)(1-P_b)},$$ (7.9)

$$F_2 = \frac{Q_b[D_b + F_1(1-P_b)] - D_b}{1-P_b}.$$ (7.10)

The calculated F_1 and F_2 for batch melting may yield fairly reliable results for primary magmas formed at low degrees of partial melting (Maaloe, 1994).

Note that if the mineral/melt partition coefficient changes in the course of melting, then we cannot obtain linear equations for Q_a and Q_b .

7.1.2. Linear regression method

The linear regression method, also called source ratio method, only works for batch melting, because the equations for fractional melting or dynamic melting cannot be expressed as linear equations. Assuming batch melting, the following linear equation can be derived (Treuil and Joron, 1975; Minster and Allegre, 1978):

$$\frac{C_a}{C_b} = \left[D_b - \frac{1-P_b}{1-P_a} D_a \right] \frac{C_a}{C_b^0} + \frac{\left(1-P_b\right)C_a^0}{\left(1-P_a\right)C_b^0} . \tag{7.11}$$

From the c_a/c_b - c_a diagrams, the values of the slope (S) and the intercept (I) can be obtained from Eq. (7.11):

$$S = \left[D_b - \frac{1-P_b}{1-P_a} D_a \right] \frac{1}{C_b^0} , \tag{7.12}$$

$$I = \frac{\left(1-P_b\right)C_a^0}{\left(1-P_a\right)C_b^0} . \tag{7.13}$$

At $c_a = 0$, if P_a and P_b are small or of similar value, then

$$I = \frac{C_a^0}{C_b^0} . \tag{7.14}$$

In this way we obtained the approximate source concentration ratio I, and then the degree of partial melting can be calculated. From Eqs. (7.5) and (7.6), the required equation is obtained as follows:

$$\frac{C_a^0}{C_b^0} = I = \frac{C_a\left[D_a + F\left(1-P_a\right)\right]}{C_b\left[D_b + F\left(1-P_b\right)\right]} . \tag{7.15}$$

Let

$$R = \frac{C_a}{C_b} , \tag{7.16}$$

based on (7.15) and (7.16), we get

$$F = \frac{RD_a - ID_b}{I\left(1-P_b\right) - R\left(1-P_a\right)} . \tag{7.17}$$

Equations (7.11) to (7.17) are the source ratio method of Treuil and Joron (1975) and Minster and Allègre (1978). The approach may be affected by the approximation at Eq. (7.14) and especially the uncertainty of the intercept I and S. In the C_a/C_b vs. C_a diagrams, the plots sometimes constitute a poor linear relationship (Clague and Frey, 1982; Giannetti and Ellam, 1994), in part because the trace element concentration deviates from the batch melting model. This may result in large errors in the intercept value I and S.

For a highly incompatible element, a, assuming $D_a = P_a = 0$, from Eqs. (7.12) and (7.13), we have:

$$S = \frac{D_b}{C_b^0}, \tag{7.18}$$

$$I = \frac{(1 - P_b)C_a^0}{C_b^0}, \tag{7.19}$$

and Eq. (7.11) becomes

$$\frac{C_a}{C_b} = C_a S + I, \tag{7.20}$$

this is, S and I are the slope and intercept of the straight line described by Eq. (7.20).

By rearranging Eq. (7.19), we obtain

$$\frac{C_b^0}{C_a^0} = \frac{1 - P_b}{I}, \tag{7.21}$$

By rearranging Eqs (7.19) and (7.20), we have

$$\frac{D_b}{C_a^0} = \frac{S(1 - P_b)}{I}. \tag{7.22}$$

Equations (7.18) to (7.22) are all from Hofmann and Feigenson (1983). The system does not obtain absolute source concentrations and partial melting degrees, but can give the source concentrations and distribution coefficients relative to the source concentration of element a.

Cebriá and López-Ruiz (1995) proposed a method to calculate mantle source concentrations, bulk distribution coefficients (D) and constant

P. They combined three Eqs. (7.6), (7.12) and (7.13) to calculate three unknowns of c_b^0, D_b, and P_b for different elements after estimating F, c_a^0, D_a, and P_a. The advantage of the method of Cebriá and López-Ruiz (1995) is that it embodies fewer assumptions about the mineralogical propositions of the mantle and is independent of the choice of D values for minerals. However, the results of c_b^0, D_b, and P_b depend on the accuracy of F and c_a^0. The F and c_a^0 values were estimated by setting $D_a = P_a = 0$, and assuming a source element concentration in the mantle with narrow limits (e.g., Yb and Lu). They established the following system of equations (Cebria and Lopez-Ruiz, 1995):

$$C_a = \frac{C_a^0}{F}, \tag{7.23}$$

$$C_{Yb} = \frac{C_{Yb}^0}{D_{Yb} + F(1 - P_{Yb})}, \tag{7.24}$$

$$S = \frac{D_{Yb}}{C_{Yb}^0}, \tag{7.25}$$

and

$$I = \frac{(1 - P_b)C_a^0}{C_{Yb}^0}. \tag{7.26}$$

The c_{Yb}^0 value is assumed to be known and c_a, c_{Yb}, S and I can be measured. The four unknowns (c_a^0, F, D_{Yb}, and P_{Yb}) can be obtained from the system of four equations [(7.23) to (7.26)]. It is evident that the obtained c_a^0 and F depend on the accuracy of c_{Yb}^0 in the mantle and S and I values. Although the c_{Yb}^0 value in the mantle varies within a narrow range, the range of 2-4*the chondritic values (Frey, 1984; McDonough and Frey, 1989) may still affect the accuracy of the c_a^0 and F values, which eventually may affect the calculated c_b^0, D_b, and P_b values. The calculated F values may not as realistic as those by the concentration ratio method for batch melting or by the equations presented here.

7.2. Dynamic Melting Inversion of Melt Compositions

Problem. Demonstrate that low-degree/high-degree concentration ratio during dynamic melting is also inversely related to the bulk partition coefficients.

The low-degree/high-degree concentration ratio in the extracted melt is

$$Q = \frac{C_1}{C_2} = \frac{X_2\left\{1-[1-X_1]^{1/[\Phi+(1-\Phi)D]}\right\}}{X_1\left\{1-[1-X_2]^{1/[\Phi+(1-\Phi)D]}\right\}}. \tag{7.27}$$

Differentiation of Q with respect to D gives

$$\frac{\partial Q}{\partial D} = -\frac{\Phi X_1 X_2}{b^2\left[\Phi+(1-\Phi)D\right]^3} \times$$

$$\left\{\begin{array}{l}\left[1-(1-X_1)^{\frac{1}{\Phi+(1-\Phi)D}}\right](1-X_2)^{\frac{1}{\Phi+(1-\Phi)D}-1} \\ +\left[1-(1-X_2)^{\frac{1}{\Phi+(1-\Phi)D}}\right](1-X_1)^{\frac{1}{\Phi+(1-\Phi)D}-1}\end{array}\right\}. \tag{7.28}$$

Since $\partial Q/\partial D$ is always negative, then Q increases with decreasing D (more incompatible).

Batch melting equations with constant mineral/melt K^i can be expressed as linear equations. In contrast, the dynamic melting equations do not display linear relationships. Therefore, the linear regression method cannot be used for dynamic melting. But we can use concentration ratio method to calculate the source concentrations and the degree of partial melting during dynamic melting.

The concentration of a trace element in the extracted dynamic melt is (Zou, 1998)

$$C = \frac{1}{X}C_0\left[1-(1-X)^{\frac{1}{\Phi+(1-\Phi)D_0}}\right], \tag{7.29}$$

where C_0 is the initial concentration of the element in the source, D is the bulk distribution coefficient, X is the mass fraction of liquid extracted relative to the initial solid, Φ is the mass porosity of the residue and is related to the volume porosity ϕ by the following relationship

$$\Phi = \frac{\rho_f \phi}{\rho_f \phi + \rho_s (1 - \phi)}, \tag{7.30}$$

where ρ_f is the density of melt ($\sim 2.8 \text{g/cm}^3$ for basaltic melt) and ρ_s is the density of solid matrix (~ 3.3 g/cm^3 for peridotite).

The enrichment ratio Q for the highly incompatible element (e.g., Th, Rb, P, and Ba) is

$$Q_a = \frac{C_a^1}{C_a^2} = \frac{X_2}{X_1} \frac{1 - (1 - X_1)^{1/[\Phi + (1-\Phi)D_a]}}{1 - (1 - X_2)^{1/[\Phi + (1-\Phi)D_a]}}. \tag{7.31}$$

Similarly, for the less incompatible element (e.g., rare earth elements (REE)),

$$Q_b = \frac{C_b^1}{C_b^2} = \frac{X_2}{X_1} \frac{1 - (1 - X_1)^{1/[\Phi + (1-\Phi)D_b]}}{1 - (1 - X_2)^{1/[\Phi + (1-\Phi)D_b]}}. \tag{7.32}$$

Equations (7.31) and (7.32) are from Zou and Zindler (1996) and Zou et al. (2000). We select both highly incompatible and not-so-highly incompatible elements because they have large but different enrichment ratios (Q) in magmas formed at different degrees of partial melting. The important feature of Q_a and Q_b is that both of them are independent of the source concentration (C_0). Equations (7.31) and (7.32) constitute a set of nonlinear equations. They have only two unknowns, X_1 and X_2, and can be solved by Newton-Raphson's method for non-linear system of equations (see Appendix 7A). The calculated F_1 and F_2 from the concentration method for batch melting using Eqs. (7.9) and (7.10) can be used as a good initial guess for the numerical solution of the system of Eqs. (7.31) and (7.32) for dynamic melting. After obtaining X_1 and X_2, the partial melting degree can be calculated by:

$$F = \Phi + (1 - \Phi)X, \tag{7.33}$$

where the first term and the second term in Eq. (7.33) represent the mass fraction of residual liquid and extracted liquid, respectively.

The source concentrations can be calculated by:

$$C_0 = \frac{\overline{C}X}{1-(1-X)^G}, \tag{7.34}$$

where

$$G = \frac{1}{\Phi + (1-\Phi)D_0}. \tag{7.35}$$

Example. Calculate the partial melting degrees and mantle source compositions for the two co-genetic basalts HS2-2 and HN6-1 from Wudalianchi, northeast China, using the dynamic melting inversion (Zou et al., 2003). The mantle critical melting porosity is constrained by U-Th disequilibrium as ϕ=0.4%. The trace element concentrations for the two co-genetic basalts HS2-2 and HN 6-1 and the bulk partition coefficients (D) are listed in Table 7.1

Table 7.1. The compositions of two co-genetic basalts.

Elements	D	HS2-2 (ppm)	HN6-1 (ppm)
La	0.0021	111.9	80.1
Nd	0.0095	80.0	59.4
Sm	0.0180	13.9	10.7
Eu	0.0228	3.75	2.93
Gd	0.0279	9.81	7.84
Tb	0.0330	1.19	0.97

For dynamic partial melting inversion, the highly incompatible element has bulk partition coefficient D_a about 0.001, and the ideal D_b for the less-so-highly incompatible element would be between 0.01 and 0.1 so that Q_b is different enough from Q_a and from 1.0. Thus, La is suitable for D_a, and Nd, Sm, Eu, Gd, and Tb are suitable for D_b. As can be seen from Table 7.2, the low-degree/high-degree Q values decrease from La (1.397) to Tb (1.227), which satisfies the requirement for inversion.

The La concentration ratio (HS2-2/HN6-1) is taken as Q_a, and the Sm concentration ratio is used to derive Q_b for this pair:

$$Q_a = Q_{La} = 111.9/80.1 = 1.397,$$

$$Q_b = Q_{Sm} = 13.9/10.7 = 1.299.$$

Using these concentration ratios and $\phi = 0.4\%$, and solving the system of Eqns. (7.31) and (7.32), we obtain $X_1 = 4.54\%$ for HS2-2 and $X_2 = 6.34\%$ for HN6-1. Substituting these values into Eq. (7.33), we have $F_1 = 4.86\%$ for HS2-2 and $F_2 = 6.65\%$ for HN6-1. Similarly, using the La concentration ratio for Q_a, but Nd, Eu, Gd, or Tb (instead of Sm) to obtain Q_b, four additional sets of F_1 and F_2 can be obtained. Averaging all the obtained values yields $X_1 = 4.62\%$, $F_1 = 5.00\%$ for HS2-2, $X_2 = 6.48\%$, and $F_2 = 6.85\%$ for HN6-1. Substituting these average values for X_1 or X_2 into Eq. (7.34), we can calculate the source concentrations (C_0) for other elements La, Nd, Sm, Eu, Gd, and Tb, respectively (Table 7.2). The results from dynamic melting inversion clearly demonstrate 4 to 7% partial melting of a mantle source that was enriched in light rare earth elements.

Table 7.2. The inversion results of the partial melting degrees and source compositions.

	D	HS2-2	HN6-1	Q	$F_1(\%)$	$F_2(\%)$	C_0	$(C_0)_N$
La	0.0021	111.9	80.1	1.397			5.17	21.9
Nd	0.0095	80.0	59.4	1.347	4.07	5.56	3.81	8.3
Sm	0.0180	13.9	10.7	1.299	4.86	6.65	0.73	4.9
Eu	0.0228	3.75	2.93	1.280	5.29	7.26	0.21	3.7
Gd	0.0279	9.81	7.84	1.251	5.35	7.35	0.59	3.0
Tb	0.0330	1.19	0.97	1.227	5.42	7.44	0.076	2.1
Ave.					5.00	6.85		

REE abundances for CI chondrites to normalize source concentration C_0 are from (Anders and Ebihara, 1982)

7.3. U-Th Disequilibrium Inversion of Melt Compositions

According to Chapter 4, when the degree of partial melting F is not too small, and melting time T is large, the U-Th disequilibrium may be described by the following relationship:

$$\frac{\left(^{230}Th\right)_{\infty}}{\left(^{238}U\right)_{\infty}} = \frac{F_{Th}}{F_U}\left(\frac{\alpha_U + \lambda_{238} + \lambda_{230} + \theta}{\alpha_{Th} + \lambda_{230} + \theta}\right),\qquad(7.36)$$

where

$$\alpha_{Th} = \frac{\left(1-D_{Th}\right)\dot{M}}{\rho_f\phi + D_{Th}\rho_s\left(1-\phi\right)},\qquad(7.37)$$

$$\alpha_U = \frac{\left(1-D_U\right)\dot{M}}{\rho_f\phi + D_U\rho_s\left(1-\phi\right)},\qquad(7.38)$$

$$F_{Th} = \frac{\rho_f\phi}{D_{Th}\rho_s\left(1-\phi\right)+\rho_f\phi},\qquad(7.39)$$

$$F_U = \frac{\rho_f\phi}{D_U\rho_s\left(1-\phi\right)+\rho_f\phi},\qquad(7.40)$$

$$\theta = \frac{\dot{M}}{\rho_s\left(1-\phi\right)}.\qquad(7.41)$$

$\left(^{230}Th\right)_{\infty}\big/\left(^{238}U\right)_{\infty}$, θ, α_{Th}, and α_U are functions of both ϕ and $\dot{M}$, whereas F_{Th} and F_U are functions of ϕ only. Note that the notations F_{Th} and F_U have nothing to do with the notations of the degree of melting. By substituting Eqs. (7.37) to (7.41) into Eq. (7.36) and after rearranging the resulting equation, we can express $\dot{M}$ as a function of ϕ for a given $(^{230}Th/^{238}U)_{\infty}$ ratio (Zou et al., 2003)

$$\dot{M} = \frac{\left(\lambda_{238}+\lambda_{230}\right)\left[\rho_f\phi + \rho_s\left(1-\phi\right)D_U\right] - Z\lambda_{230}\left[\rho_f\phi + \rho_s\left(1-\phi\right)D_{Th}\right]}{\left(Z-1\right)\left[1+\rho_f\phi\big/\left[\rho_s\left(1-\phi\right)\right]\right]},$$

$$(7.42)$$

where

$$Z = (^{230}Th/^{238}U)_{\infty}.\qquad(7.43)$$

Alternatively, we can express ϕ as a function of $\dot{M}$, but Eq (7.42) is simpler. There are two unknowns in ϕ and $\dot{M}$ but only one equation (7.42), and we cannot obtain two parameters from one equation. Nevertheless, we can put some quantitative constraints on both ϕ and $\dot{M}$ from $(^{230}Th / ^{238}U)_\infty$.

Example. Estimate the maximum melting rate and the maximum porosity for lavas with $(^{230}Th/^{238}U)=1.25$ and $(^{230}Th/^{238}U)=1.33$, respectively. The bulk partition coefficients are $D_{Th}=0.003$ and $D_U=0.005$, for Th and U, respectively; the densities are $\rho_f=2800$ g/cm^3, $\rho_s=3300$ g/cm^3 for melt and solid respectively; and the decay constants are $\lambda_{230Th}=9.217*10^{-6}$ year^{-1} and $\lambda_{238U}=1.55125*10^{-10}$ year^{-1}.

We at first try to make a graph (Fig. 7.1) to find the maximum melting rate ($\dot{M}_{max}$) and the maximum porosity (ϕ_{max}).

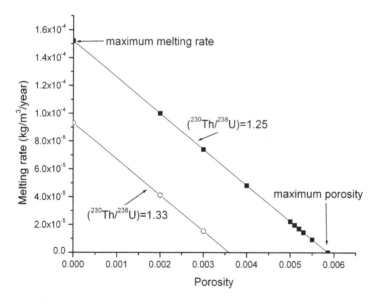

Fig. 7.1. Inversion of U-Th disequilibrium generated by partial melting. The maximum melting rate and maximum porosity can be obtained from the graph based on measured $(^{230}Th/^{238}U)$ values.

Based on Eq. (7.42), and using the given parameters, we can make a melting rate ($\dot{M}$) vs. porosity (ϕ) figure (Fig. 7.1) that satisfy this relationship (7.42) for a given value of (^{230}Th/^{238}U). We know that neither $\dot{M}$ nor ϕ can be negative. For melting rates to be non-negative, the melting porosity ϕ must be <0.6%. For non-negative values of ϕ, the melting rate $\dot{M}$ must be $<1.6\times10^{-4}$ (kg/m^3/year) (Fig. 7.1).

The maximum melting rate ($\dot{M}_{max}$) and the maximum porosity (ϕ_{max}) can also be directly calculated. When $\phi \to 0$, the melting rate is at the maximum and Eq. (7.42) can be reduced to

$$\dot{M}_{max} = \frac{(\lambda_{238} + \lambda_{230})\rho_s D_U - Z\lambda_{230}\rho_s D_{Th}}{Z-1}. \tag{7.44}$$

Similarly, the porosity is at its maximum when $\dot{M} \to 0$ in Eq. (7.42), and can be calculated by

$$\phi_{max} = \frac{Z\lambda_{230}(\rho_f - \rho_s D_{Th}) - (\lambda_{238} + \lambda_{230})(\rho_f - \rho_s D_U)}{(\lambda_{238} + \lambda_{230})\rho_s D_U - Z\lambda_{230}\rho_s D_{Th}}. \tag{7.45}$$

Note that Z is the measured (^{230}Th/^{238}U) value.

7.4. Batch Melting Inversion of Whole-rock Residues

7.4.1. Concentration ratio method

In the context of non-modal (eutectic) batch melting, the variation in the concentration of a trace element in a whole-rock residue, C_r, is given by Shaw (1970) as

$$C_r = \frac{C_0}{1-F} \frac{D_0 - PF}{D_0 + F(1-P)},$$

where C^0 is the concentration in the source. Let the degree of partial melting increase from stage 1 to stage 2 ($F_2 > F_1$). The concentration of a highly incompatible element (e.g., La) in the residue varies from C_{ra}^1 to C_{ra}^2, where

$$C_{ra}^1 = \frac{C_a^0}{1 - F_1} \frac{D_a^0 - P_a F_1}{D_a^0 + F_1(1 - P_a)},$$

and

$$C_{ra}^2 = \frac{C_a^0}{1 - F_2} \frac{D_a^0 - P_a F_2}{D_a^0 + F_2(1 - P_a)}.$$

The concentration ratio for the highly incompatible element is given by

$$R_a = \frac{C_{ra}^1}{C_{ra}^2} = \frac{(1 - F_2)(D_a^0 - P_a F_1)\left[D_a^0 + F_2(1 - P_a)\right]}{(1 - F_1)(D_a^0 - P_a F_2)\left[D_a^0 + F_1(1 - P_a)\right]}. \tag{7.46}$$

Similarly, the concentration ratio for the less incompatible element is defined by

$$R_b = \frac{C_{rb}^1}{C_{rb}^2} = \frac{(1 - F_2)(D_b^0 - P_b F_1)\left[D_b^0 + F_2(1 - P_b)\right]}{(1 - F_1)(D_b^0 - P_b F_2)\left[D_b^0 + F_1(1 - P_b)\right]}. \tag{7.47}$$

Both highly incompatible and not-so-highly incompatible elements are selected because they have large but different concentration ratios in residues formed after different degrees of partial melting. The important feature of R_a and R_b is that both are independent of the source concentration (C^0). Equations (7.46) and (7.47) constitute a system of nonlinear equations and do not yield analytical solutions. The two unknowns, F_1 and F_2, can be solved by Newton's method for a system of non-linear equations. After obtaining F_1 and F_2, the source concentrations can be calculated from the relationship:

$$C^0 = \frac{C_r(1 - F)\left[D^0 + F(1 - P)\right]}{D^0 - PF}. \tag{7.48}$$

However, if $D_a^0 = P_a$ and $D_b^0 = P_b$ in Eqs. (7.46) and (7.47), which is the case for modal batch melting, the system yields analytical solutions (Zou, 1997)

$$F_1 = \frac{D_a^0(1 - D_b^0)(1 - R_a) + D_b^0(1 - D_a^0)(R_b - 1)}{(R_a - R_b)(1 - D_a^0)(1 - D_b^0)}, \tag{7.49}$$

$$F_2 = \frac{R_b\left[D_b^0 + F_1(1-D_b^0)\right] - D_b^0}{1-D_b^0}.$$ (7.50)

7.4.2. Linear regression method

The linear regression method, or slope-intercept method (Zou, 1997), for whole-rock residues can only be applied to modal batch melting. For elements a (highly incompatible) and b (not-so-highly incompatible), their concentrations in the residue during modal batch melting, C_{ra} and C_{rb}, respectively, are given by

$$C_{ra} = \frac{D_a^0 C_a^0}{D_a^0 + F(1-D_a^0)},$$ (7.51)

and

$$C_{rb} = \frac{D_b^0 C_b^0}{D_b^0 + F(1-D_b^0)}.$$ (7.52)

The following relationship between C_{ra}/C_{rb} and C_{ra} can be derived from Eqs. (7.51) and (7.52):

$$\frac{C_{ra}}{C_{rb}} = \left[D_b^0 - \frac{D_a^0(1-D_b^0)}{1-D_a^0}\right]\frac{C_{ra}}{D_b^0 C_b^0} + \frac{D_a^0(1-D_b^0)C_a^0}{D_b^0(1-D_a^0)C_b^0}.$$ (7.53)

Equation (7.53) shows that the ratio C_{ra}/C_{rb}, is a linear function of the concentration C_{ra}. The slope (S) and intercept (I) of the straight line described by Eq. (7.53) can be expressed as

$$S = \left[D_b^0 - \frac{D_a^0(1-D_b^0)}{1-D_a^0}\right]\frac{1}{D_b^0 C_b^0},$$

$$I = \frac{D_a^0(1-D_b^0)C_a^0}{D_b^0(1-D_a^0)C_b^0}.$$

We can obtain S and I by linear regression from the C_{ra}/C_{rb} vs. C_{ra} diagram if there is a significant linear relationship between C_{ra}/C_{rb} and C_{ra} for a set of peridotites. Source concentrations can be calculated by

$$C_a^0 = \frac{I}{S}\left[\frac{D_b^0\left(1-D_a^0\right)}{D_a^0\left(1-D_b^0\right)}-1\right],$$

and

$$C_b^0 = \frac{C_a^0}{SC_a^0+I}.$$

Alternatively, a linear relationship between $1/C_{ra}$ and $1/C_{rb}$ for modal batch melting can also be obtained from Eqs. (7.51) and (7.52) as follows:

$$\frac{1}{C_{rb}} = S\frac{1}{C_{ra}}+I, \tag{7.54}$$

where the slope (S) and the intercept (I) of the straight line described by Eq. (7.54) can be expressed as

$$S = \frac{D_a^0\left(1-D_b^0\right)C_a^0}{D_b^0\left(1-D_a^0\right)C_b^0},$$

$$I = \left[D_b^0 - \frac{D_a^0\left(1-D_b^0\right)}{1-D_a^0}\right]\frac{1}{D_b^0 C_b^0}.$$

After obtaining S and I in the $1/C_{ra}$ vs. $1/C_{rb}$ diagram by linear regression, source concentrations are given

$$C_a^0 = \frac{S}{I}\left[\frac{D_b^0\left(1-D_a^0\right)}{D_a^0\left(1-D_b^0\right)}-1\right],$$

$$C_b^0 = \frac{C_a^0}{IC_a^0+S}.$$

7.5. Batch Melting Inversion of Residual Clinopyroxenes

If residual peridotites are pervasively serpentinized, the whole-rock residues would be unsuitable for inversion of partial melting. In most cases clinopyroxene (cpx) is unaltered, and cpx concentration ratios can be used with confidence in constraining the magmatic processes affecting the rock as a whole. Clinopyroxene contains the highest concentrations of incompatible trace elements in typical peridotites (olivine + clinopyroxene + orthopyroxene + spinel), and can be measured in thin sections of serpentinized peridotites using the ion microprobe (Johnson et al., 1990).

7.5.1. Concentration ratio method

Assuming the cpx distribution coefficient is a constant ($D^{cpx} = D^{0,cpx}$), the relationship between C_r / C^0 and $C^{cpx} / C^{0,cpx}$ is given by Johnson et al. (1990) as

$$\frac{C_r}{C^0} = \frac{C^{cpx}}{C^{0,cpx}} \frac{D}{D^0} . \tag{7.55}$$

The basic equations describing the variation in concentration of a trace element in cpx of non-modal batch melting model can be obtained as follows (Johnson et al., 1990):

$$C^{cpx} = C^{0,cpx} \frac{D^0}{D^0 + F(1-P)} . \tag{7.56}$$

By using the similar approach as mentioned in section 7.4.1, we can obtain a system of equations consisting of the cpx concentration ratio for the highly incompatible element (R_a^{cpx}) and the ratio for the less incompatible element (R_b^{cpx}) from Eq. (7.56).

R_a^{cpx} is given by

$$R_a^{cpx} = \frac{C_{cpx,a}^1}{C_{cpx,a}^2} = \frac{D_a^0 + F_2\left(1-P_a\right)}{D_a^0 + F_1\left(1-P_a\right)} . \tag{7.57}$$

R_b^{cpx} is defined by

$$R_b^{cpx} = \frac{C_{Cpx,b}^1}{C_{Cpx,b}^2} = \frac{D_b^0 + F_2\left(1-P_b\right)}{D_b^0 + F_1\left(1-P_b\right)} . \tag{7.58}$$

The solutions to the system of simultaneous equations (7.57) and (7.58) are (Zou, 1997)

$$F_1 = \frac{D_a^0(1-P_b)(1-R_a^{cpx}) + D_b^0(1-P_a)(R_b^{cpx}-1)}{(R_a^{cpx} - R_b^{cpx})(1-P_a)(1-P_b)}, \quad (7.59)$$

$$F_2 = \frac{R_b^{cpx}\left[D_b^0 + F_1(1-P_b)\right] - D_b^0}{1-P_b}. \quad (7.60)$$

After obtaining F, the source cpx concentration ($C^{0,cpx}$) for each model can be calculated by substituting F into Eq. (7.56). The relationship between the source whole-rock concentration and the source cpx concentration is

$$C^0 = C^{0,cpx}\frac{D^0}{D^{cpx}}. \quad (7.61)$$

7.5.2. Linear regression method

In contrast to the case for whole rocks, the linear regression method for residual clinopyroxenes can be applied to non-modal batch melting. The basic equation for the variation of a trace element in cpx during non-modal batch melting has been given in Eq. (7.56). We can obtain the following linear relationship from Eq. (7.56) (Zou, 1997):

$$\frac{C_a^{cpx}}{C_b^{cpx}} = SC_a^{cpx} + I, \quad (7.62)$$

where

$$S = \left[D_b^0 - \frac{D_a^0(1-P_b)}{1-P_a}\right]\frac{1}{D_b^0 C_b^{0,cpx}}, \quad (7.63)$$

$$I = \frac{D_a^0(1-P_b)C_a^{0,cpx}}{D_b^0(1-P_a)C_b^{0,cpx}}. \quad (7.64)$$

After obtaining S and I from the plot of C_a^{cpx}/C_b^{cpx} vs. C_a^{cpx}, the source cpx concentrations can be obtained by

$$C_a^{0,cpx} = \frac{I}{S}\left[\frac{D_b^0\left(1-P_a\right)}{D_a^0\left(1-P_b\right)}-1\right], \qquad (7.65)$$

$$C_b^{0,cpx} = \frac{C_a^{0,cpx}}{SC_a^{0,cpx}+I}. \qquad (7.66)$$

Alternatively, the following linear relationship can be obtained from Eq. (7.56):

$$\frac{1}{C_b^{cpx}} = S\frac{1}{C_a^{cpx}}+I, \qquad (7.67)$$

where

$$S = \frac{D_a^0\left(1-P_b\right)C_a^{0,cpx}}{D_b^0\left(1-P_a\right)C_b^{0,cpx}}, \qquad (7.68)$$

$$I = \left[D_b^0 - \frac{D_a^0\left(1-P_b\right)}{1-P_a}\right]\frac{1}{D_b^0 C_b^{0,cpx}}. \qquad (7.69)$$

After obtaining S and I in the $1/C_a^{cpx}$ vs. $1/C_b^{cpx}$ diagram by linear regression, source concentrations are given by

$$C_a^{0,cpx} = \frac{S}{I}\left[\frac{D_b^0\left(1-P_a\right)}{D_a^0\left(1-P_b\right)}-1\right],$$

$$C_b^{0,cpx} = \frac{C_a^{0,cpx}}{IC_a^{0,cpx}+S}.$$

Appendix 7A. Newton-Raphson Method for Two Unknowns

We wish to solve the nonlinear system of two equations:

$$F(x,y)=0,$$

$$G(x,y)=0,$$

where are given functions of the independent variables x and y. After choosing an approximate solution (x_0, y_0), i.e., an initial approximation, we set

$$x = x_0 + \varepsilon_0,$$

$$y = y_0 + \tau_0,$$

and expand the two functions F and G in Taylor series about the point (x_0, y_0). Since ($x_0 + \varepsilon_0$, $y_0 + \tau_0$) is a solution of the system of equations, the results are

$$0 = F(x_0 + \varepsilon_0, y_0 + \tau_0) = F(x_0, y_0) + \frac{\partial F}{\partial x}\varepsilon_0 + \frac{\partial F}{\partial y}\tau_0 + \text{higher order terms},$$

$$0 = G(x_0 + \varepsilon_0, y_0 + \tau_0) = G(x_0, y_0) + \frac{\partial G}{\partial x}\varepsilon_0 + \frac{\partial G}{\partial y}\tau_0 + \text{higher order terms},$$

or, in matrix form:

$$\begin{vmatrix} 0 \\ 0 \end{vmatrix} = \begin{vmatrix} F \\ G \end{vmatrix}_{(x_0,y_0)} + \begin{vmatrix} \partial F/\partial x & \partial F/\partial y \\ \partial G/\partial x & \partial G/\partial y \end{vmatrix}_{(x_0,y_0)} \begin{vmatrix} \varepsilon_0 \\ \tau_0 \end{vmatrix} + \text{higher order terms}.$$

Assuming the existence of

$$\begin{vmatrix} \partial F/\partial x & \partial F/\partial y \\ \partial G/\partial x & \partial G/\partial y \end{vmatrix}^{-1},$$

we have

$$\begin{vmatrix} \varepsilon_0 \\ \tau_0 \end{vmatrix} = \begin{vmatrix} \partial F/\partial x & \partial F/\partial y \\ \partial G/\partial x & \partial G/\partial y \end{vmatrix}^{-1}_{(x_0,y_0)} \begin{vmatrix} F \\ G \end{vmatrix}_{(x_0,y_0)} + \text{higher order terms},$$

which gives as a solution (x,y):

$$\begin{vmatrix} x \\ y \end{vmatrix} = \begin{vmatrix} x_0 \\ y_0 \end{vmatrix} - \begin{vmatrix} \partial F/\partial x & \partial F/\partial y \\ \partial G/\partial x & \partial G/\partial y \end{vmatrix}^{-1}_{(x_0,y_0)} \begin{vmatrix} F \\ G \end{vmatrix}_{(x_0,y_0)} + \text{higher order terms}.$$

When we neglect the higher order terms, the solution (x, y) becomes an approximation, which will be taken as the new starting point (x_1, y_1) for the next approximation:

$$\begin{vmatrix} x_1 \\ y_1 \end{vmatrix} = \begin{vmatrix} x_0 \\ y_0 \end{vmatrix} - \begin{vmatrix} \partial F/\partial x & \partial F/\partial y \\ \partial G/\partial x & \partial G/\partial y \end{vmatrix}_{(x_0,y_0)}^{-1} \begin{vmatrix} F \\ G \end{vmatrix}_{(x_0,y_0)}.$$

Thus, we find the recurring formula of Newton-Raphson for two unknowns:

$$\begin{vmatrix} x_{k+1} \\ y_{k+1} \end{vmatrix} = \begin{vmatrix} x_k \\ y_k \end{vmatrix} - \begin{vmatrix} \partial F/\partial x & \partial F/\partial y \\ \partial G/\partial x & \partial G/\partial y \end{vmatrix}_{(x_k,y_k)}^{-1} \begin{vmatrix} F \\ G \end{vmatrix}_{(x_k,y_k)}, \quad k=0,\ 1,2,...$$

References

Allegre, C.J. and Minster, J.F., 1978. Quantitative models of trace-element behavior in magmatic processes. Earth Planet. Sci. Lett., 38.

Anders, E. and Ebihara, M., 1982. Solar-system abundances of the elements. Geochim. Cosmochim. Acta, 46: 2363-2380.

Cebria, J.M. and Lopez-Ruiz, J., 1995. Alkali basalts and leucitites in an extensional intracontinental plate setting: The late Cenozoic Calatrava Volcanic Province (central Spain). Lithos, 35: 27-46.

Clague, D.A. and Frey, F.A., 1982. Petrology and trace element geochemistry of the Honolulu volcanics, Oahu: Implications for the oceanic mantle below Hawaii. J. Petrol., 23: 447-504.

Class, C. and Goldstein, S.L., 1997. Plume-lithosphere interactions in the ocean basin: constraints from the source mineralogy. Earth Planet. Sci. Lett., 150: 245-260.

Frey, F.A., 1984. Rare earth element abundances in the upper mantle rocks. In: P. Henderson (Editor), Rare Earth Elelemtn Geochemistry. Elsevier, Amsterdam, pp. 153-203.

Frey, F.A., Green, D.H. and Roy, S.D., 1978. Integrated models of basalts petrogenesis: a study of quartz tholeiite to olivine melilitites from South Eastern Australia utilizing geochemical and experimental data. J. Petrol., 19: 463-513.

Giannetti, B. and Ellam, R., 1994. The primitive lavas of Roccamonfina volcano, Roman region, Italy: new constraints on melting processes and source mineralogy. Contrib. Mineral. Petrol., 116: 21-31.

Hofmann, A.W. and Feigenson, M.D., 1983. Case studies on the origin of basalt: I. Theory and reassessment of Grenada basalts. Contrib. Mineral. Petrol., 84: 382-389.

Johnson, K.T.M., Dick, H.J.B. and Shimizu, N., 1990. Melting in the oceanic mantle: An ion microprobe study of diopside in abyssal peridotite. J. Geophys. Res., 95: 2661-2678.

Maaloe, S., 1994. Estimation of the degree of partial melting using concentration ratios. Geochim. Cosmochim. Acta, 58: 2519-2525.

McDonough, W.F. and Frey, F.A., 1989. Rare earth elements in the upper mantle rocks. In: B.R. Lipin and G.A. McKay (Editors), Geochemistry and Mineralogy of Rare Earth Elements. Mineral. Soc. Amer.

McKenzie, D. and O'Nions, R.K., 1991. Partial melt distributions from inversion of rare earth element concentrations. J. Petrol., 32: 1021-1091.

Minster, J.F. and Allegre, C.J., 1978. Systematic use of trace elements in igneous processes. Part III: Inverse problem of batch partial melting in volcanic suites. Contrib. Mineral. Petrol., 68: 37-52.

Shaw, D.M., 1970. Trace element fractionation during anatexis. Geochim. Cosmochim. Acta, 34: 237-243.

Treuil, L. and Joron, J.L., 1975. Utilisation des elements hygromagmatophiles pour la simplification de la modelisation quantitative des processus magmatique. Examples de l'Afar et la dorsale medioatlantique. Soc. Ital. Mineral. Petrol., 31: 125-174.

Zou, H.B., 1997. Inversion of partial melting through residual peridotites or clinopyroxenes. Geochim. Cosmochim. Acta, 61(21): 4571-4582.

Zou, H.B., 1998. Trace element fractionation during modal and nonmodal dynamic melting and open-system melting: A mathematical treatment. Geochim. Cosmochim. Acta, 62: 1937-1945.

Zou, H.B., Reid, M.R., Liu, Y.S., Yao, Y.P., Xu, X.S. and Fan, Q.C., 2003. Constraints on the origin of historic potassic basalts from northeast China by U-Th disequilibrium data. Chem. Geol., 200: 189-201.

Zou, H.B. and Zindler, A., 1996. Constraints on the degree of dynamic partial melting and source composition using concentration ratios in magmas. Geochim. Cosmochim. Acta, 60: 711-717.

Zou, H.B., Zindler, A., Xu, X.S. and Qi, Q., 2000. Major and trace element, and Nd-Sr-Pb isotope studies of Cenozoic basalts in SE China: mantle sources, regional variations, and tectonic significance. Chem. Geol., 171: 33-47.

Chapter 8

Error Analysis

Estimate of errors is fundamental to all branches of natural sciences that deal with experiments. Very frequently, the final result of an experiment cannot be measured directly. Rather, the value of the final result (u) will be calculated from several measured quantities ($x, y, z...,$, each of which has a mean value and an error):

$$u = f(x, y, z...) .$$ (8.1)

The goal is to estimate the error in the final result u from the errors in measured quantities x, y, z. The errors can be random or systematic. Random errors displace measurements in an arbitrary direction whereas systematic errors displace measurements in a single direction. Systematic errors shift all measurement in a systematic way so that their mean value is displaced. For example, a miscalibrated ruler may yield consistently higher values of length. Random errors fluctuate from one measurement to the next. They yield results distributed about the same mean value. Random errors can be treated by statistical analysis based on repeated measurements whereas systematic errors cannot.

8.1. Random Errors

Both random errors and systematic errors in the measurements of x, y, or z lead to error in the determination of u in Eq. (8.1). Since systematic errors shift measurement in a single direction, we use du to treat systematic errors. In contrast, since random errors can be both positive and negative, we use $(du)^2$ to treat random errors.

$$\left(du\right)^2 = \left(\frac{\partial u}{\partial x}\right)^2_{y,z}\left(dx\right)^2 + \left(\frac{\partial u}{\partial y}\right)^2_{x,z}\left(dy\right)^2 + \left(\frac{\partial u}{\partial z}\right)^2_{x,y}\left(dz\right)^2$$

$$+2\left(\frac{\partial u}{\partial x}\right)_{y,z}\left(\frac{\partial u}{\partial y}\right)_{x,z}dxdy + 2\left(\frac{\partial u}{\partial y}\right)_{x,z}\left(\frac{\partial u}{\partial z}\right)_{x,y}dydz$$

$$+2\left(\frac{\partial u}{\partial x}\right)_{y,z}\left(\frac{\partial u}{\partial z}\right)_{x,y}dxdz.$$

$$(8.2)$$

If the measured variables are independent (non-correlated), then the cross-terms average to zero

$$dxdy = 0,\ dydz = 0,\ dxdz = 0,\qquad(8.3)$$

as dx, dy, and dz each take on both positive and negative values.

Thus,

$$du = \sqrt{\left(du\right)^2}$$

$$= \sqrt{\left(\frac{\partial u}{\partial x}\right)^2 dx^2 + \left(\frac{\partial u}{\partial y}\right)^2 dy^2 + \left(\frac{\partial u}{\partial z}\right)^2 dz^2}.\qquad(8.4)$$

Equating standard deviation with differential, *i.e.*,

$$\sigma_u = du,\ \sigma_x = dx,\ \sigma_y = dy,\ \text{and}\ \sigma_z = dz,$$

results in the error propagation formula for independent (non-correlated) random errors

$$\sigma_u = \sqrt{\left(\frac{\partial u}{\partial x}\right)^2_{y,z}\sigma_x^{\ 2} + \left(\frac{\partial u}{\partial y}\right)^2_{x,z}\sigma_y^{\ 2} + \left(\frac{\partial u}{\partial z}\right)^2_{x,y}\sigma_z^{\ 2}}.\qquad(8.5)$$

Equation (8.5) for independent random errors has been frequently used in the uncertainty analysis of many physical and chemical experiments.

Example. Calculate the independent error in the volume of a rectangle resulting from 0.1 cm random error in the measured x, y, z, when $x = 10$ cm, $y = 20$ cm, and $z = 5$ cm.

Since the volume of a rectangle is $V = xyz$, we have the partial derivatives as

$$\left(\frac{\partial V}{\partial x}\right)_{y,z} = yz, \quad \left(\frac{\partial V}{\partial y}\right)_{x,z} = xz, \quad \left(\frac{\partial V}{\partial z}\right)_{x,y} = xy.$$

The errors are $\sigma_x = \sigma_y = \sigma_z = 0.1$ cm, thus we have

$$\sigma_V = \sqrt{\left(\frac{\partial V}{\partial x}\right)_{y,z}^2 \sigma_x^2 + \left(\frac{\partial V}{\partial y}\right)_{x,z}^2 \sigma_y^2 + \left(\frac{\partial V}{\partial z}\right)_{x,y}^2 \sigma_z^2}$$

$$= \sqrt{(yz)^2 \sigma_x^2 + (xz)^2 \sigma_y^2 + (xy)^2 \sigma_z^2}$$

$$= \sqrt{(10 \times 5)^2 0.1^2 + (20 \times 5)^2 0.1^2 + (20 \times 10)^2 0.1^2}$$

$$= 23 \text{ cm}^3.$$

Therefore, the expected uncertainty in V is 23 cm^3. The relative error is $s_V / V = 23 / 1000 = 2.3\%$.

In contrast, if we have +0.1 cm systematic error, then the uncertainty is

$$du = \left(\frac{\partial u}{\partial x}\right)_{y,z} dx + \left(\frac{\partial u}{\partial y}\right)_{x,z} dy + \left(\frac{\partial u}{\partial z}\right)_{x,y} dz$$

$$= (10 \times 5) \times 0.1 + (20 \times 5) \times 0.1 + (20 \times 10) \times 0.1$$

$$= +35 \text{ cm}^3.$$

Note that the random error (23 cm^3) is less than the systematic error (35 cm^3) in this case.

Example. Calculate the error in $(^{230}\text{Th}/^{238}\text{U})$ from the errors in $(^{230}\text{Th}/^{232}\text{Th})$ and $(^{238}\text{U}/^{232}\text{Th})$.

In isotope geochemistry, $(^{230}\text{Th}/^{238}\text{U})$ reflects U-Th disequilibrium. This ratio is not measured directly from mass spectrometers. In stead, isotope composition measurement provides $(^{230}\text{Th}/^{232}\text{Th})$ and isotope dilution measurement gives (^{238}U/^{232}Th). Calculate the error in (^{230}Th/^{238}U) when (^{230}Th/^{232}Th) =1.464±0.011 and (^{238}U/^{232}Th)=1.296±0.013.

Let x=(^{230}Th/^{232}Th) and y=(^{238}U/^{232}Th), then

$$u = (^{230}Th/^{238}U) = x/y = 1.130 .$$

The partial derivatives are

$$\frac{\partial u}{\partial x} = 1/y ,$$

$$\frac{\partial u}{\partial y} = -x/y^2 .$$

Thus, the uncertainty in U is

$$\sigma_U = \sqrt{\left(\frac{\partial u}{\partial x}\right)_y^2 \sigma_x^2 + \left(\frac{\partial u}{\partial y}\right)_{x,z}^2 \sigma_y^2}$$

$$= \sqrt{(1/y)^2 \sigma_x^2 + (-x/y^2)^2 \sigma_y^2}$$

$$= \sqrt{(1/1.464)^2 0.013^2 + (-1.296/1.464^2)^2 0.011^2}$$

$$= 0.011 .$$

Therefore, the result is

$$u = (^{230}Th/^{238}U) = 1.130 \pm 0.011 .$$

Note that, during mass spectrometric analyses, the isotopic composition (^{230}Th/^{232}Th) is measured independently from (^{238}U/^{232}Th), and thus the errors of these two ratios are not correlated, even though they have the same denominator in ^{232}Th. (^{238}U/^{232}Th) is obtained by isotope dilution method while (^{230}Th/^{232}Th) is measured directly by mass spectrometers.

Example. Estimate the total volume of a mixture from three batches: $x = 30 \pm 0.5$ ml, $y = 40 \pm 0.4$ ml, and $z = 50 \pm 0.6$ ml.

The partial derivatives for $u = x + y + z$ are

$$\left(\frac{\partial u}{\partial x}\right)_{y,z} = \left(\frac{\partial u}{\partial y}\right)_{x,z} = \left(\frac{\partial u}{\partial z}\right)_{x,y} = 1 .$$

And the uncertainty in u is

$$\sigma_u = \sqrt{\left(\frac{\partial u}{\partial x}\right)_{y,z} \sigma_x^2 + \left(\frac{\partial u}{\partial y}\right)_{x,z} \sigma_y^2 + \left(\frac{\partial u}{\partial z}\right)_{y,z} \sigma_z^2}$$

$$= \sqrt{\sigma_x^2 + \sigma_y^2 + \sigma_z^2} = \sqrt{0.5^2 + 0.4^2 + 0.6^2}$$

$$= 0.88 \approx 0.9 \text{ ml.}$$

Therefore we have

$$u = 120.0 \pm 0.9 \text{ ml}.$$

Example. To make 120 ml solution, assuming the error for each batch is the same (for example, 0.5 ml), calculate the error (1) we mix two batches of 60 ml solution, and (2) if we mix three batches of 40 ml solution.

For the first scenario, we use 60 ml batches twice, each with 0.5 ml error.

$$\sigma_u = \sqrt{\left(\frac{\partial u}{\partial x}\right)_{y,z} \sigma_x^2 + \left(\frac{\partial u}{\partial y}\right)_{y,z} \sigma_y^2}$$

$$= \sqrt{\sigma_x^2 + \sigma_y^2} = \sqrt{0.5^2 + 0.5^2}$$

$$= 0.7 \text{ ml.}$$

Therefore, the result is

$$u = 120.0 \pm 0.7 \text{ ml}.$$

For the second scenario, we mix three batches of 40 ml solution, each with 0.5 ml error. The error in u is

$$\sigma_u = \sqrt{\left(\frac{\partial u}{\partial x}\right)_{y,z} \sigma_x^2 + \left(\frac{\partial u}{\partial y}\right)_{y,z} \sigma_y^2 + \left(\frac{\partial u}{\partial z}\right)_{y,z} \sigma_z^2}$$

$$= \sqrt{\sigma_x^2 + \sigma_y^2 + \sigma_z^2} = \sqrt{0.5^2 + 0.5^2 + 0.5^2}$$

$$= 0.86 \approx 0.9 \text{ ml.}$$

Therefore we have

$$u = 120.0 \pm 0.9 \text{ ml}.$$

Thus, if each batch has the same error, it is better to reduce the number of batches in order to reduce the total error.

Example. For the second scenario (3 batches of 40 ml solution), in order to have the same total error (0.7 ml) as the first scenario, what is the error for each measurement?

For three batches with total error of 0.7 ml, we have

$$\sigma_u = \sqrt{\sigma_x^2 + \sigma_y^2 + \sigma_z^2} = \sqrt{3\sigma_x^2} = 0.7 \text{ ml}.$$

The solution is $\sigma_x = 0.4 \text{ ml}$. It means that, in order to achieve the same total error, we have to be more precise each time for the second scenario.

Example. Determine the principal source of error to make cylinder with 20 cm diameter and 5 cm high and suggest improvement.
The volume of the cylinder is

$$u = \pi r^2 h = \pi \times 10^2 \times 5 = 1570 \text{ cm}^3.$$

The partial derivatives are

$$\left(\frac{\partial u}{\partial r}\right)_h = 2\pi rh = 2\pi \times 10 \times 5 = 314 \text{ cm}^2,$$

$$\left(\frac{\partial u}{\partial h}\right)_r = \pi r^2 = \pi 10^2 = 314 \text{ cm}^2.$$

Since $\left(\frac{\partial u}{\partial r}\right)_h = \left(\frac{\partial u}{\partial h}\right)_r$, both r and h are equally critical in this case. If error is 0.2 cm for both r and h, we have

$$
\begin{aligned}
\sigma_U &= \sqrt{\left(\frac{\partial u}{\partial r}\right)_h^2 \sigma_r^2 + \left(\frac{\partial u}{\partial h}\right)_r^2 \sigma_h^2} \\
&= \sqrt{(2\pi rh)^2 \sigma_r^2 + (\pi r^2)^2 \sigma_h^2} \\
&= \sqrt{(2\pi \times 10 \times 5)^2 0.2^2 + (\pi \times 10^2)^2 0.2^2} \\
&= 89 \text{ cm}^3.
\end{aligned}
$$

The volume of the cylinder is

$$u = 1570 \pm 89 \ \text{cm}^3 .$$

Question. What if it is a cylinder with 20 cm diameter but 20 cm height?

$$\left(\frac{\partial u}{\partial r} \right)_h = 2\pi r h = 2\pi \times 10 \times 20 = 1256 \ \text{cm}^2 ,$$

$$\left(\frac{\partial u}{\partial h} \right)_r = \pi r^2 = \pi 10^2 = 314 \ \text{cm}^2 .$$

Since $\left(\dfrac{\partial u}{\partial r} \right)_h > \left(\dfrac{\partial u}{\partial h} \right)_r$, for this long (20 cm high) cylinder, the precision of r is more critical.

Question. What if it is a very flat cylinder with 20 cm diameter but 3 cm height?

$$\left(\frac{\partial u}{\partial r} \right)_h = 2\pi r h = 2\pi \times 10 \times 3 = 188 \ \text{cm}^2 ,$$

$$\left(\frac{\partial u}{\partial h} \right)_r = \pi r^2 = \pi 10^2 = 314 \ \text{cm}^2 .$$

Since $\left(\dfrac{\partial u}{\partial h} \right)_r > \left(\dfrac{\partial u}{\partial r} \right)_h$, for this short (3 cm high) cylinder, the precision of h is more critical.

Example. Calculate the uncertainty of the acceleration of gravity measured with a simple pendulum based on the length of a pendulum and the period of the pendulum. The acceleration of gravity g is related to the length l and the period T by the following relationship:

$$g = 4\pi^2 l / T^2 .$$

The measured $l = 99.43 \pm 0.08$ cm and the measured $T = 2.000 \pm 0.003$ second. Based on these parameters, we get

$$g = 4\pi^2 l / T^2 = 4\pi^2 \times 99.43 / 2^2 = 981.33 \ \text{cm/s}^2 .$$

The partial derivative of g with respect to l is

$$\left(\frac{\partial g}{\partial l}\right)_T = \frac{4\pi^2}{T^2} = \frac{g}{l}.$$

And the partial derivative of g with respect to T is

$$\left(\frac{\partial g}{\partial T}\right)_l = -\frac{8\pi^2 l}{T^3} = -\frac{2g}{T}.$$

The absolute uncertainty of g is

$$\sigma_g = \sqrt{\left(\left(\frac{\partial g}{\partial l}\right)_T \sigma_l\right)^2 + \left(\left(\frac{\partial g}{\partial T}\right)_l \sigma_T\right)^2} = g\sqrt{\left(\frac{\sigma_l}{l}\right)^2 + \left(-\frac{2\sigma_T}{T}\right)^2}$$

$$= 981.33\sqrt{\left(\frac{0.008}{99.43}\right)^2 + \left(-\frac{2\times 0.003}{2.000}\right)^2}$$

$$= 2.95 \text{ cm/s}^2.$$

Therefore, the measured acceleration of gravity g is

$$g = 981 \pm 3 \text{ cm/s}^2.$$

The relative uncertainty is

$$\sigma_g/g = 2.95/981.33 = 0.3\%.$$

So far we have been using the general Eq. (8.5) for the propagation of independent random errors. For some special cases, we may use simpler equations. If u is the sum or difference of x, y, and z, then according to Eq. (8.5), we obtain

$$\sigma_u = \sqrt{\sigma_x^2 + \sigma_y^2 + \sigma_z^2}. \tag{8.6}$$

If u is the product or quotient of x, y, and z, then

$$\left(\frac{\sigma_u}{u}\right)^2 = \left(\frac{\sigma_x}{x}\right)^2 + \left(\frac{\sigma_y}{y}\right)^2 + \left(\frac{\sigma_z}{z}\right)^2. \tag{8.7}$$

If u is a power, $u = x^n$, where n is known exactly, then

$$\frac{\sigma_u}{|u|} = |n| \frac{\sigma_x}{|x|} .$$ (8.8)

8.2. Statistical Treatment of Random Uncertainties

Random errors can be treated by statistical analysis based on repeated measurements whereas systematic errors cannot.

8.2.1. Mean and standard deviation

Suppose we make N measurements of the quantity x (using the identical equipment and procedures) and find the N values x_1, x_2, ... x_N. The mean of the measurements is

$$\bar{x} = \frac{\sum_{1}^{N} x_i}{N} .$$ (8.9)

Standard deviation (σ_x) is

$$\sigma_x = \sqrt{\frac{1}{N-1} \sum_{1}^{N} (x_i - \bar{x})^2} .$$ (8.10)

If we make one more measurement of x then it would have some 68% probability of lying within $\bar{x} \pm \sigma_x$. The standard deviation represents the average uncertainty in the individual measurements x_1, x_2, ... x_N. Thus, if we want to make some more measurements using the same equipment and procedures, the standard deviation would not change significantly (e.g., Taylor, 1997). In contrast, the standard error, or the standard deviation of the mean, would surely decrease as we increase the number of the measurements N. Standard error ($\sigma_{\bar{x}}$), or, standard deviation of the mean, is given by

$$\sigma_{\bar{x}} = \frac{\sigma_x}{\sqrt{N}} .$$ (8.11)

The meaning of the standard error is that if the N measurements of x are repeated, there would be a 68% probability that the new mean value would lie within $\bar{x} \pm \sigma_{\bar{x}}$, and there would be a 95% probability that the new mean value would lie within $\bar{x} \pm 2\sigma_{\bar{x}}$.

Problem. Prove Eq. (8.11) for the standard deviation of the mean. Equation (8.9) can be rewritten as

$$\bar{x} = \frac{x_1 + x_2 + \ldots + x_N}{N}. \tag{8.12}$$

Differentiation of $\bar{x}$ with respect to $x_1, x_2, \ldots, x_i, \ldots, x_N$, respectively, yields

$$\frac{\partial \bar{x}}{\partial x_1} = \frac{\partial \bar{x}}{\partial x_2} = \ldots = \frac{\partial \bar{x}}{\partial x_i} = \ldots = \frac{\partial \bar{x}}{\partial x_N} = \frac{1}{N}. \tag{8.13}$$

From the error propagation principle, we obtain

$$\sigma_{\bar{x}} = \sqrt{\left(\frac{\partial \bar{x}}{\partial x_1}\sigma_{x_1}\right)^2 + \ldots + \left(\frac{\partial \bar{x}}{\partial x_i}\sigma_{x_i}\right)^2 + \ldots + \left(\frac{\partial \bar{x}}{\partial x_n}\sigma_{x_N}\right)^2}$$

$$= \sqrt{\left(\frac{\sigma_{x_1}}{N}\right)^2 \ldots + \left(\frac{\sigma_{x_i}}{N}\right)^2 + \ldots + \left(\frac{\sigma_{x_N}}{N}\right)^2}. \tag{8.14}$$

Since

$$\sigma_{x_1} = \sigma_{x_i} = \sigma_{x_N} = \sigma_x, \tag{8.15}$$

we get

$$\sigma_{\bar{x}} = \sqrt{\left(\frac{\sigma_x}{N}\right)^2 \ldots + \left(\frac{\sigma_x}{N}\right)^2 + \ldots + \left(\frac{\sigma_x}{N}\right)^2} = \sqrt{\frac{N\sigma_x^2}{N^2}}$$

$$= \frac{\sigma_x}{\sqrt{N}}. \tag{8.16}$$

8.2.2. Weighted averages

We often need to find the best estimate of a quantity from two or more separate and independent measurements of a single physical or chemical quantity. Suppose we have N separate measurements of a quantity x,

$$x_1 \pm \sigma_1, \ x_2 \pm \sigma_2, \ x_3 \pm \sigma_3, \ ..., \ x_N \pm \sigma_N,$$

with their corresponding uncertainties σ_1, σ_2, σ_3, ..., σ_N. The best estimate based on these measurements is the weighted average

$$\bar{x}_\omega = \frac{\sum \omega_i x_i}{\sum \omega_i}, \tag{8.17}$$

and the weight of each measurement is the reciprocal square of the corresponding uncertainty

$$\omega_i = 1/\sigma^2. \tag{8.18}$$

Because the weight associated with each measurement involves the square of the corresponding uncertainty, the measurement that is more precise than others contributes very much more to the final weighted average.

The uncertainty in x_ω is

$$\sigma_\omega = 1/\sqrt{\sum \omega_i}. \tag{8.19}$$

Problem. Prove Eq. (8.19) for the standard deviation of the weighted mean.

Equation (8.17) can be written as

$$\bar{x}_\omega = \frac{\omega_1 x_1 + \omega_2 x_2 + ... + \omega_i x_i + ... + \omega_N x_N}{\sum \omega_i}. \tag{8.20}$$

Differentiation of Eq. (8.20) with respect to x_1, x_2, x_i, x_N, respectively, gives

$$\frac{\partial \bar{x}_\omega}{\partial x_1} = \frac{\omega_1}{\sum \omega_i}, \tag{8.21}$$

$$\frac{\partial \overline{x}_\omega}{\partial x_2} = \frac{\omega_2}{\sum \omega_i}, \tag{8.22}$$

$$\frac{\partial \overline{x}_\omega}{\partial x_i} = \frac{\omega_i}{\sum \omega_i}, \tag{8.23}$$

$$\frac{\partial \overline{x}_\omega}{\partial x_N} = \frac{\omega_N}{\sum \omega_i}. \tag{8.24}$$

Substitution of (8.21), (8.22), (8.23), (8.24) into Eq. (8.5) for error propagation results in

$$\sigma_{\overline{x}_\omega} = \sqrt{\left(\frac{\partial \overline{x}_\omega}{\partial x_1}\sigma_{x_1}\right)^2 + \dots + \left(\frac{\partial \overline{x}_\omega}{\partial x_i}\sigma_{x_i}\right)^2 + \dots + \left(\frac{\partial \overline{x}_\omega}{\partial x_n}\sigma_{x_N}\right)^2}$$

$$= \sqrt{\left(\frac{\omega_1\sigma_{x_1}}{\sum \omega_i}\right)^2 \dots + \left(\frac{\omega_i\sigma_{x_i}}{\sum \omega_i}\right)^2 + \dots + \left(\frac{\omega_N\sigma_{x_N}}{\sum \omega_i}\right)^2}. \tag{8.25}$$

Since

$$\omega_i = 1/\sigma_i^2, \tag{8.26}$$

we obtain

$$\sigma_{\overline{x}_\omega} = \sqrt{\frac{\omega_1}{\left(\sum \omega_i\right)^2} \dots + \frac{\omega_i}{\left(\sum \omega_i\right)^2} + \dots + \frac{\omega_N}{\left(\sum \omega_i\right)^2}}$$

$$= 1/\sqrt{\sum \omega_i}. \tag{8.27}$$

Example. Repeated ion microprobe (e.g., Cameca IMS 1270) measurements of a large homogeneous zircon provide the following individual ages (in Ma, or million years):

$$t_1 = 1105 \pm 35, \ t_2 = 1195 \pm 60,$$

$$t_3 = 1125 \pm 20, \ t_4 = 1150 \pm 25,$$

$$t_5 = 1095 \pm 45.$$

Given these results, calculate the best estimate for the zircon age using weighted average.

The five uncertainties are 35, 60, 20, 25 and 45. Therefore, the corresponding weights $\omega_i = 1/\sigma^2$ are

$$\omega_1 = 1/35^2 = 0.00082, \quad \omega_2 = 1/60^2 = 0.00028,$$

$$\omega_3 = 1/20^2 = 0.0025, \quad \omega_4 = 1/25^2 = 0.0016,$$

$$\omega_5 = 1/45^2 = 0.00049.$$

Thus we obtain

$$\sum \omega_i = 0.00082 + 0.00028 + 0.0025 + 0.0016 + 0.00049$$
$$= 0.00569,$$

$$\sum \omega_i t_i = 1105 \times 0.00082 + 1195 \times 0.00028 + 1125 \times 0.0025$$
$$+ 1150 \times 0.0016 + 1095 \times 0.00049$$
$$= 6.42722.$$

The above calculation of ω_i, $\sum \omega_i$ and $\sum \omega_i t_i$ are summarized in Table 8.1.

Table 8.1. Calculation of the weighted zircon age and uncertainty.

	t	σ_T	ω_i	$\omega_i t_i$
	1105	35	0.00082	0.90204
	1195	60	0.00028	0.33194
	1125	20	0.0025	2.81250
	1150	25	0.0016	1.84000
	1095	45	0.00049	0.54074
Sum			$\sum \omega_i$	$\sum \omega_i t_i$
			$= 0.00569$	$= 6.42722$
t_ω	1130			
σ_ω		13		

The best estimate is the weighted average

$$T_\omega = \frac{\sum \omega_i t_i}{\sum \omega_i} = \frac{6.42722}{0.00569} = 1130 \text{ Ma.}$$

And the uncertainty is

$$\sigma_\omega = 1\Big/\sqrt{\sum \omega_i} = 1/\sqrt{0.00569} = 13 \text{ Ma.}$$

Thus, the final age for the zircon is

$$T_\omega = 1130 \pm 13 \text{ Ma.}$$

Example. Multiple in-situ measurements of oxygen isotopic compositions in a single olivine grain give the following values (in per mil):

$$x_1 = 6.2 \pm 0.7, \ x_2 = 6.3 \pm 0.4,$$

$$x_3 = 6.9 \pm 0.5, \ x_4 = 6.4 \pm 0.3,$$

$$x_5 = 6.1 \pm 0.6, \ x_6 = 6.8 \pm 0.8.$$

The six uncertainties are 0.7, 0.4, 0.5, 0.3, 0.6 and 0.8. Therefore, the corresponding weights $\omega_i = 1/\sigma^2$ are

$$\omega_1 = 1/0.7^2 = 2.04, \ \omega_2 = 1/0.4^2 = 6.25,$$

$$\omega_3 = 1/0.5^2 = 4, \ \omega_4 = 1/0.3^2 = 11.11,$$

$$\omega_5 = 1/0.6^2 = 2.78, \ \omega_6 = 1/0.8^2 = 1.56.$$

Based on the above parameters, we get

$$\sum \omega_i = 2.04 + 6.25 + 4.00 + 11.11 + 2.78 + 1.56 = 27.74,$$

$$\sum \omega_i T_i = 6.2 \times 2.04 + 6.3 \times 6.25 + 6.9 \times 4.00 + 6.4 \times 11.11$$
$$+ 6.1 \times 2.78 + 6.8 \times 1.56$$
$$= 178.31.$$

The best estimate is the weighted average

$$x_\omega = \frac{\sum \omega_i x_i}{\sum \omega_i} = \frac{178.31}{27.74} = 6.43.$$

And the uncertainty is

$$\sigma_\omega = 1 \Big/ \sqrt{\sum \omega_i} = 1/\sqrt{27.74} = 0.19.$$

Thus, the oxygen isotopic composition for the olivine is

$$x_\omega = 6.4 \pm 0.2 \text{ per mil.}$$

The above calculation can be summarized in Table 8.2.

Table 8.2. Calculation of weighted oxygen isotope compositions and uncertainty.

	x_i	σ_{x_i}	ω_i	$\omega_i x_i$
	6.2	0.7	2.04	12.65
	6.3	0.4	6.25	39.38
	6.9	0.5	4.00	27.60
	6.4	0.3	11.11	71.11
	6.1	0.6	2.78	16.94
	6.8	0.8	1.56	10.625
Sum			$\sum \omega_i = 27.74$	$\sum \omega_i x_i = 178.31$
x_ω	6.43			
σ_ω		1.19		

8.2.3. Variance, covariance and correlated errors

The variance for N times of measurement of x_i is

$$\sigma_x^2 = \frac{1}{N} \sum_{i=1}^{N} (x_i - \overline{x}_i)^2. \tag{8.28}$$

The variance for N times of measurement of y_i is

$$\sigma_y^2 = \frac{1}{N} \sum_{i=1}^{N} (y_i - \overline{y}_i)^2. \tag{8.29}$$

And the covariance is

$$\sigma_{xy} = \frac{1}{N} \sum_{i=1}^{N} \left(x_i - \bar{x}_i \right) \left(y_i - \bar{y}_i \right). \qquad (8.30)$$

u is dependent on x and y by

$$u = u(x, y).$$

The error in u is

$$\sigma_u^2 = \left(\frac{\partial u}{\partial x} \right)^2 \sigma_x^2 + \left(\frac{\partial u}{\partial y} \right)^2 \sigma_y^2 + 2 \frac{\partial u}{\partial x} \frac{\partial u}{\partial y} \sigma_{xy}. \qquad (8.31)$$

For independent errors, $\sigma_{xy} = 0$, and Eq. (8.31) reduces to

$$\sigma_u^2 = \left(\frac{\partial u}{\partial x} \right)^2 \sigma_x^2 + \left(\frac{\partial u}{\partial y} \right)^2 \sigma_y^2. \qquad (8.32)$$

Example. Suppose we have the following five times of measurements of K_2O and Na_2O:

   first measurement: $K_2O = 3.0$, $Na_2O = 1.69$;
   second measurement: $K_2O = 3.1$, $Na_2O = 1.80$;
   third measurement: $K_2O = 2.9$, $Na_2O = 1.60$;
   fourth measurement: $K_2O = 3.3$, $Na_2O = 1.85$;
   fifth measurement: $K_2O = 2.8$, $Na_2O = 1.55$.

Calculate the best estimate of uncertainty of the total alkali ($Na_2O + K_2O$) using Eq. (8.31).

Let $x = K_2O$, $y = Na_2O$.

The averages of x and y can be obtained as

$$\bar{x} = \overline{K_2O} = 3.02$$

$$\bar{y} = \overline{Na_2O} = 1.698.$$

Thus $\bar{u} = \bar{x} + \bar{y} = 3.02 + 1.698 = 4.718$.

The individual values of $(x_i - \bar{x})$, $(y_i - \bar{y})$, $(x_i - \bar{x})(y_i - \bar{y})$ are given in Table 8.3.

Table 8.3. Calculation of uncertainty of total alkali (Na_2O+K_2O).

	x_i	y_i	$(x_i - \bar{x})$	$(y_i - \bar{y})$	$(x_i - \bar{x})(y_i - \bar{y})$
1	3.0	1.69	-0.02	-0.008	0.00016
2	3.1	1.80	0.08	0.102	0.00816
3	2.9	1.60	-0.12	-0.098	0.01176
4	3.3	1.85	0.28	0.152	0.04256
5	2.8	1.55	-0.22	-0.148	0.03256

The variance of x_i is

$$\sigma_x^2 = \frac{1}{N}\sum_{i=1}^{N}(x_i - \bar{x}_i)^2$$
$$= \frac{1}{5}\left[(-0.02)^2 + 0.08^2 + (-0.12)^2 + 0.28^2 + (-0.22)^2\right]$$
$$= 0.0296.$$

Thus we have $\sigma_x = 0.17$.

The variance of y_i is

$$\sigma_y^2 = \frac{1}{N}\sum_{i=1}^{N}(y_i - \bar{y})^2$$
$$= \frac{1}{5}\left[(-0.008)^2 + 0.102^2 + (-0.098)^2 + 0.152^2 + (-0.148)^2\right]$$
$$= 0.01302.$$

Thus we have $\sigma_y = 0.11$.

Note that the values of $(x_i - \bar{x})(y_i - \bar{y})$ in Table 8.3 are all positive, resulting in positive covariance of x_i and y_i.
The covariance of x_i and y_i is

$$\sigma_{xy} = \frac{1}{N}\sum_{i=1}^{N}(x_i - \bar{x}_i)(y_i - \bar{y}_i)$$
$$= \frac{1}{5}\left[\begin{array}{l}(-0.02)(-0.008)+0.08\times0.102+(-0.12)(-0.098) \\ +0.28\times0.152+(-0.22)(-0.148)\end{array}\right]$$
$$= 0.019.$$

For $u = x + y$, the partial derivatives are

$$\frac{\partial u}{\partial x} = 1 \text{ and } \frac{\partial u}{\partial y} = 1.$$

From Eq. (8.31) we get

$$\sigma_u = \sqrt{\left(\frac{\partial u}{\partial x}\right)^2 \sigma_x^2 + \left(\frac{\partial u}{\partial y}\right)^2 \sigma_y^2 + 2\frac{\partial u}{\partial x}\frac{\partial u}{\partial y}\sigma_{xy}}$$

$$= \sqrt{\sigma_x^2 + \sigma_y^2 + 2\sigma_{xy}}$$

$$= \sqrt{0.0296 + 0.01302 + 2 \times 0.019}$$

$$= 0.28.$$

When correlated errors are considered, we have

$$\bar{u} = 4.72 \pm 0.28 .$$

If we ignore the correlated errors, then from (8.32), we have

$$\sigma_u = \sqrt{\left(\frac{\partial u}{\partial x}\right)^2 \sigma_x^2 + \left(\frac{\partial u}{\partial y}\right)^2 \sigma_y^2}$$

$$= \sqrt{\sigma_x^2 + \sigma_y^2} = \sqrt{0.0296 + 0.01302}$$

$$= 0.21.$$

When correlated errors are taken into account, because the errors are positively correlated, the resulting uncertainty (0.28) is larger than the corresponding uncertainty calculated from un-correlated error (0.21). The next example will show negatively correlated errors.

Example. Suppose we have a different data set as follows:

first measurement: $K_2O = 3.0$, $Na_2O = 1.71$;
second measurement: $K_2O = 3.1$, $Na_2O = 1.60$;
third measurement: $K_2O = 2.9$, $Na_2O = 1.80$;
fourth measurement: $K_2O = 3.3$, $Na_2O = 1.55$;
fifth measurement: $K_2O = 2.8$, $Na_2O = 1.85$.

Calculate the correlated error.
Let $x = K_2O$, $y = Na_2O$.
The averages of x and y can be obtained as

$$\bar{x} = \overline{K_2O} = 3.02, \quad \bar{y} = \overline{Na_2O} = 1.702.$$

Thus

$$\bar{u} = \bar{x} + \bar{y} = 3.02 + 1.702 = 4.722.$$

The individual values of $(x_i - \bar{x})$, $(y_i - \bar{y})$, $(x_i - \bar{x})(y_i - \bar{y})$ are given in Table 8.4.

Table 8.4. Calculation of uncertainty of total alkali (Na_2O+K_2O)

	x_i	y_i	$(x_i - \bar{x})$	$(y_i - \bar{y})$	$(x_i - \bar{x})(y_i - \bar{y})$
1	3.0	1.71	-0.02	0.012	-0.00024
2	3.1	1.60	0.08	-0.098	-0.00784
3	2.9	1.80	-0.12	0.102	-0.01224
4	3.3	1.55	0.28	-0.148	-0.04144
5	2.8	1.85	-0.22	0.152	-0.03344

The variance of x_i is

$$\sigma_x^2 = \frac{1}{N}\sum_{i=1}^{N}(x_i - \bar{x}_i)^2$$
$$= \frac{1}{5}\left[(-0.02)^2 + 0.08^2 + (-0.12)^2 + 0.28^2 + (-0.22)^2\right]$$
$$= 0.0296.$$

Thus we have $\sigma_x = 0.11$.

The variance of y_i is

$$\sigma_y^2 = \frac{1}{N}\sum_{i=1}^{N}(y_i - \bar{y}_i)^2$$
$$= \frac{1}{5}\left[0.012^2 + (-0.098)^2 + 0.102^2 + (-0.148)^2 + 0.152^2\right]$$
$$= 0.01303.$$

Thus we have $\sigma_y = 0.11$.

Note that the values of $(x_i - \bar{x})(y_i - \bar{y})$ in Table 8.4 are all negative, resulting in a negative covariance of x_i and y_i. The covariance is

$$\sigma_{xy} = \frac{1}{N}\sum_{i=1}^{N}(x_i - \overline{x}_i)(y_i - \overline{y}_i)$$

$$= \frac{1}{5}\left[\begin{array}{l}(-0.02)\times 0.012 + 0.08\times(-0.098) + (-0.12)\times 0.102 \\ +0.28\times(-0.148) + (-0.22)\times 0.152\end{array}\right]$$

$$= -0.019.$$

From (8.31) we obtain

$$\sigma_u = \sqrt{\sigma_x^2 + \sigma_y^2 + 2\sigma_{xy}}$$

$$= \sqrt{0.0296 + 0.01303 + 2\times(-0.019)}$$

$$= 0.07.$$

Therefore, we have $u = 4.722 \pm 0.07$.

For this example, when correlated errors are taken into account, the resulting uncertainty (0.07) is smaller than un-correlated errors (0.21), because the errors are negatively correlated here.

8.3. Probability Distributions

8.3.1. Normal/Gauss distribution

The normal/Gauss distribution is given by

$$G_{X,\sigma}(x) = \frac{1}{\sigma\sqrt{2\pi}} e^{-(x-X)^2/2\sigma^2}, \qquad (8.33)$$

where X is the center and σ is the width of the normal distribution. Note that this distribution is symmetrical at X. Figure 8.1 shows the significance of the width σ. A small σ gives a sharply peaked distribution (high precision) whereas a large σ gives a broad distribution (low precision).

The integral

$$\int_a^b G_{X,\sigma}(x)dx, \qquad (8.34)$$

is the probability that any one measurement gives an number in the range of $a \le x \le b$ for the Gauss distribution.

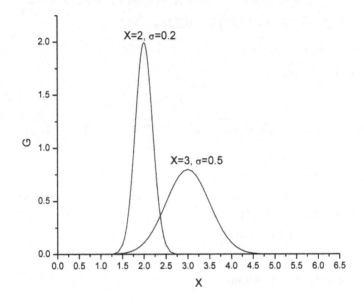

Fig. 8.1. Plot of Gauss distribution for $x = 2$, $\sigma = 0.2$ and for $x = 3$, $\sigma = 0.5$.

The probability that a measurement number will fall with one standard deviation of the true value X is thus

$$P = \int_{X+\sigma}^{X+\sigma} G_{X,\sigma}(x)dx = \frac{1}{\sigma\sqrt{2\pi}} \int_{X+\sigma}^{X+\sigma} e^{-(x-X)^2/2\sigma^2} dx \,. \qquad (8.35)$$

Let $t = (x - X)/\sigma$, thus, $dx = \sigma dt$, Eq. (8.35) can be written as

$$P = \frac{1}{\sqrt{2\pi}} \int_{-1}^{1} e^{-t^2/2} dt \,. \qquad (8.36)$$

This probability integral (8.36) can be written using the error function of mathematical physics, denoted

$$erf(z) = \frac{2}{\sqrt{\pi}} \int_{0}^{z} e^{-u^2} du \,. \qquad (8.37)$$

The error function is also useful for solving chemical diffusion problems (Chapter 13) and thermal conduction problems and the details of the error function and the table of the error function are given in Chapter 13.

Let $u = t/\sqrt{2}$ so $du = dt/\sqrt{2}$. Then Eq. (8.35) becomes

$$P = \frac{1}{\sqrt{2\pi}} \int_{-1}^{1} e^{-t^2/2} dt = \sqrt{\frac{2}{\pi}} \int_{0}^{1} e^{-t^2/2} dt = \frac{2}{\sqrt{\pi}} \int_{0}^{1/\sqrt{2}} e^{-u^2} du, \quad (8.38)$$

which is an error function when $z = 1/\sqrt{2}$. Therefore, for 1σ, the probability is

$$P(\text{within } 1\sigma) = erf\left(1/\sqrt{2}\right) = erf(0.7071). \quad (8.39)$$

From Table 13.1 of $erf(z)$ function (Chapter 13) for $z = 0.7071$, we have

$$P(\text{within } 1\sigma) = 68\%. \quad (8.40)$$

Therefore, there would be a 68% probability that a new measurement number would lie within $X \pm \sigma$.

Similarly, we can derive the probability of an answer within $n\sigma$ of X:

$$P = \frac{1}{\sqrt{2\pi}} \int_{-n}^{n} e^{-t^2/2} dt = \frac{2}{\sqrt{\pi}} \int_{0}^{n/\sqrt{2}} e^{-u^2} du = erf\left(n/\sqrt{2}\right). \quad (8.41)$$

From (8.41), when $n = 2$, we have

$$P(\text{within } 2\sigma) = erf(2/\sqrt{2}) = erf(1.414). \quad (8.42)$$

From Table 13.1 of $erf(z)$ function for $z = 1.414$, we find

$$P(\text{within } 2\sigma) = 95\%. \quad (8.43)$$

Therefore, there would be a 95% probability that the new value would lie within $X \pm 2\sigma$.

When $n = 3$, we have

$$P(\text{within } 3\sigma) = erf(3/\sqrt{2}) = erf(2.121). \quad (8.44)$$

From Table 13.1 of $erf(z)$ function for $z = 2.121$, we find

$$P(\text{within } 3\sigma) = 99.7\%. \quad (8.45)$$

Therefore, there would be a 99.7% probability that the new value would lie within $X \pm 3\sigma$.

8.3.2. Poisson distribution

The Poisson distribution describes the results of experiments in which we count events that occur at random but at a definite average rate. Examples of the Poisson distribution include the number of emails we receive in a one-day period, the number of babies born in a hospital in a two-day period, the number of decays of a radioactive isotope in a one-day period.

The Poisson distribution is described by:

$$P_\mu(v) = e^{-\mu} \frac{\mu^v}{v!} \,. \tag{8.46}$$

The Poisson distribution gives the probability of getting the result v in an experiment in which we count events that occur at random but a definite average rate μ. The standard deviation of the Poisson distribution with mean count μ is $\sqrt{\mu}$:

$$\sigma_P = \sqrt{\mu} \,. \tag{8.47}$$

Example. Suppose we know that we receive five emails (excluding junk emails) as an average in a one-day period, what is the probability that we receive three emails in a specific day?
In this example, we have $\mu = 5$ and $v = 3$. According to Eq. (8.46), we get the probability of receiving three emails in one day as

$$P_5(3) = e^{-5} \frac{5^3}{3!} = 14\% \,.$$

Example. For the same average (5 emails/day), what is the chance of receiving 10 emails in a specific day?
In this case, we have $\mu = 5$ and $v = 10$. And the chance is

$$P_5(10) = e^{-5} \frac{5^{10}}{10!} = 1.8\% \,.$$

Figure 8.2 displays the Poisson distribution for $\mu = 5$ and Fig. 8.3 illustrates the Poisson distribution for $\mu = 3$.

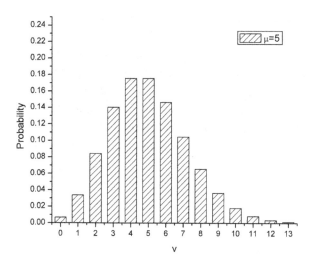

Fig. 8.2. Poisson distribution for $\mu = 5$.

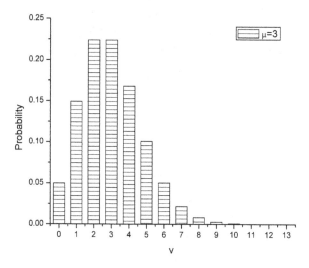

Fig. 8.3. Poisson distribution for $\mu = 3$.

The Poisson distribution has important applications in isotope geochemistry where counting statistics is needed. One example in radioactive isotope geochemistry is the counting of the natural decays of a radioactive nuclide, e.g., ^{226}Ra, in a sample, by alpha-counting methods. Another example is the counting of ions generated by thermal ionization mass spectrometer, secondary ion mass spectrometer (e.g., ion microprobe), or inductively-coupled mass spectrometer. For small ion beams, we often need to use ion counters, for example, Daly ion counters or electron multipliers, to count ions generated by mass spectrometers (large ion beams are measured by Faraday cups which are not related to counting statistics).

We always want to estimate the measurement errors from ion counting. If we count 10,000 ions using ion counters, then from (8.47), the error from counting statistics would be

$$\sigma_P = \sqrt{10000} = 100 .$$

And the relative error is

$$\sigma_P / \mu = 100/10000 = 1\% .$$

If we want a better relative error, we need to increase the counting time and/or the intensity of the ion beam. For example, to achieve a relative error of 0.1%, then from (8.47) we have

$$\sqrt{\mu}/\mu = 0.1\% .$$

Therefore we need $\mu = 1,000,000$ counts to achieve counting statistics error of 0.1%.

8.3.3. Binomial distribution

If the possibility of success in any one trial is p, then the possibility of v successes in n trials is given by the binomial distribution:

$$B_{n,p}(v) = \frac{n!}{v!(n-p)!} p^v (1-p)^{n-v} , \qquad (8.48)$$

where

$$\frac{n!}{v!(n-v)!} = \frac{n(n-1)\cdots(n-v+1)\times(n-v)!}{1\times2\times\cdots\times v\times(n-v)!}$$
$$= \frac{n(n-1)\cdots(n-v+1)}{1\times2\times\cdots\times v}, \tag{8.49}$$

is called the binomial coefficient. Note that $n! = 1\times2\times\cdots\times n$. The binomial coefficient appears in the binomial expansion

$$(p+q)^n = p^n + np^{n-1}q + \cdots + q^n$$
$$= \sum_{v=0}^{n}\binom{n}{v}p^v q^{n-v}. \tag{8.50}$$

If we repeat the whole set of n trials many times, the expected mean number of successes is

$$\overline{v} = np, \tag{8.51}$$

and the standard deviation of v is

$$\sigma_v = \sqrt{np(1-p)}. \tag{8.52}$$

8.4. Systematic Errors

A systematic error in the measurement of x, y, or z leads to an error in the determination of u:

$$du = \left(\frac{\partial u}{\partial x}\right)_{y,z} dx + \left(\frac{\partial u}{\partial y}\right)_{x,z} dy + \left(\frac{\partial u}{\partial z}\right)_{x,y} dz. \tag{8.53}$$

This is essentially the multi-dimensional definition of slope. It describes how changes in u depend on changes in x, y, and z. Note that we use du to examine systematic errors but $(du)^2$ to examine random errors.

Example. Calculate the systematic error in the volume of a cylinder resulting from a mis-calibrated ruler. A mis-calibrated ruler results in a systematic error by +0.2 cm in diameter and height are 20 cm, and 5 cm, respectively. The values of r and h in the cylinder must be changed by +0.2 cm.

Volume of a cylinder:

$$V = \pi r^2 h = \pi \times 10^2 \times 5 = 1570 \text{ cm}^3.$$

The systematic error in the volume is

$$dV = \left(\frac{\partial V}{\partial r}\right)_h dr + \left(\frac{\partial V}{\partial h}\right)_r dh$$
$$= (2\pi rh)dr + (\pi r^2)dh$$
$$= (2\pi \times 10 \times 5)(0.2) + (\pi \times 10^2)(0.2)$$
$$= +125.6 \text{ cm}^3.$$

The relative systematic error is

$$dV/V = 125.6/1570 = 8\%.$$

Example. Suppose that we have systematic errors for La concentrations in two samples with the following systematic errors as a result of miscalibrated standard.

$$x = 15 + 0.3,$$
$$y = 10 + 0.3.$$

Calculate the systematic error of concentration ratio x/y.

$$Q = x/y = 15.0/10.0 = 1.50,$$

$$\left(\frac{\partial Q}{\partial x}\right)_y = 1/y = 1/10 = 0.10,$$

$$\left(\frac{\partial Q}{\partial y}\right)_x = -x/y^2 = -15/10^2 = -0.15.$$

The systematic error in x/y is

$$dQ = \left(\frac{\partial Q}{\partial x}\right)_y dx + \left(\frac{\partial Q}{\partial y}\right)_x dy$$
$$= 0.10 \times 0.3 - 0.15 * 0.3$$
$$= -0.015,$$

and the relative systematic error is

$$dQ/Q = -0.015/0.15 = -1\% .$$

Note that the systematic error in x/y is better then the precision of both x (0.3/15=2%) and y (0.3/10=3%). This is why geochemists like to use ratios instead of concentrations, as ratios can reduce systematic errors.

For further reading of error analysis, the following books are useful: Beers (1957), Baird (1988), Bevington and Robinson (1992), and Taylor (1997).

8.5. Summary

1) *Random errors*

Correlated random errors:

$$\sigma_u^2 = \left(\frac{\partial u}{\partial x}\right)_{y,z}^2 \sigma_x^2 + \left(\frac{\partial u}{\partial y}\right)_{x,z}^2 \sigma_y^2 + \left(\frac{\partial u}{\partial z}\right)_{x,y}^2 \sigma_z^2$$

$$+ 2\left(\frac{\partial u}{\partial x}\right)_{y,z}\left(\frac{\partial u}{\partial y}\right)_{x,z} \sigma_x\sigma_y + 2\left(\frac{\partial u}{\partial y}\right)_{x,z}\left(\frac{\partial u}{\partial z}\right)_{x,y} \sigma_y\sigma_z$$

$$+ 2\left(\frac{\partial u}{\partial x}\right)_{y,z}\left(\frac{\partial u}{\partial z}\right)_{x,y} \sigma_x\sigma_z .$$

Independent random errors:

$$\sigma_u^2 = \left(\frac{\partial u}{\partial x}\right)_{y,z}^2 \sigma_x^2 + \left(\frac{\partial u}{\partial y}\right)_{x,z}^2 \sigma_y^2 + \left(\frac{\partial u}{\partial z}\right)_{x,y}^2 \sigma_z^2 .$$

Three special cases of the independent random errors: if u is the sum or difference of x, y, and z, then

$$\sigma_u = \sqrt{\sigma_x^2 + \sigma_y^2 + \sigma_z^2} ;$$

if u is the product or quotient of x, y, and z, then

$$\left(\frac{\sigma_u}{u}\right)^2 = \left(\frac{\sigma_x}{x}\right)^2 + \left(\frac{\sigma_y}{y}\right)^2 + \left(\frac{\sigma_z}{z}\right)^2 ;$$

if u is a power, $u = x^n$, where n is known exactly, then

$$\frac{\sigma_u}{|u|} = |n|\frac{\sigma_x}{|x|} .$$

2) Statistical treatment of random uncertainties

The mean of N measurements is

$$\bar{x} = \frac{x_1 + x_2 + ... + x_N}{N} = \frac{\sum_1^N x_i}{N} .$$

Standard deviation (σ_x) is

$$\sigma_x = \sqrt{\frac{1}{N-1}\sum_1^N (x_i - \bar{x})^2} .$$

Standard error ($\sigma_{\bar{x}}$), or, standard deviation of the mean, is

$$\sigma_{\bar{x}} = \frac{\sigma_x}{\sqrt{N}} .$$

The variance for N times of measurement of x_i is

$$\sigma_x^2 = \frac{1}{N}\sum_{i=1}^N (x_i - \bar{x}_i)^2 .$$

The variance for N times of measurement of y_i is

$$\sigma_y^2 = \frac{1}{N}\sum_{i=1}^N (y_i - \bar{y}_i)^2 .$$

And the covariance is

$$\sigma_{xy} = \frac{1}{N} \sum_{i=1}^{N} \left(x_i - \bar{x}_i \right) \left(y_i - \bar{y}_i \right).$$

Weighted average is

$$\bar{x}_\omega = \frac{\sum \omega_i x_i}{\sum \omega_i},$$

and the weight of each measurement is the reciprocal square of the corresponding uncertainty

$$\omega_i = 1/\sigma^2.$$

The uncertainty in x_ω is

$$\sigma_\omega = 1/\sqrt{\sum \omega_i}.$$

3) Probability distribution

Normal/Gauss distribution:

$$G_{X,\sigma}(x) = \frac{1}{\sigma\sqrt{2\pi}} e^{-(x-X)^2/2\sigma^2}.$$

The probability of an answer within $n\sigma$ of X:

$$P = erf\left(n/\sqrt{2} \right).$$

$P = 68\%$ when $n = 1$; $P = 95\%$ when $n = 2$; and $P = 99.7\%$ when $n = 3$.

Poisson distribution:

$$P_\mu(v) = e^{-\mu} \frac{\mu^v}{v!} \text{ with error } \sigma_P = \sqrt{\mu}.$$

Binomial distribution:

$$B_{n,p}(v) = \frac{n!}{v!(n-p)!} p^v (1-p)^{n-v}.$$

4) Systematic Errors

$$du_{sys} = \left(\frac{\partial u}{\partial x}\right)_{y,z} dx + \left(\frac{\partial u}{\partial y}\right)_{x,z} dy + \left(\frac{\partial u}{\partial z}\right)_{x,y} dz.$$

References

Beers, Y. (1957) Introduction to the theory of error. Addison-Wesley. Reading. 66 pp.

Baird, D. C. (1988) Experimentation: an introduction to measurement theory and experiment design. Prentice Hall. Englewood Cliffs. 193 pp.

Bevington, P. R. and Robinson, D. K. (1992) Data reduction and error analysis for the physical sciences. McGraw-Hill. New York, 328 pp.

Taylor, J. R. (1997) An introduction to error analysis. University Science Books. Sausalito. 327 pp.

Chapter 9

Linear Least Square Fitting

Linear least square fitting is a mathematical procedure for finding the best-fitting line $y_i = a + bx_i$ to a given set of points (X_i, Y_i) by minimizing the sum of the squares of the residuals. This method has wide application in almost every branch of natural sciences and over one hundred articles have been published in a variety of journals in the fields of physics, applied mathematics, earth sciences, chemistry and biology (e.g., Adcock, 1878; Pearson, 1901; Bartlett, 1949; Kermack and Haldane, 1950; Deming, 1964; McIntyre et al., 1966; York, 1966; York, 1967; York, 1969; Lybanon, 1984; Reed, 1989; Cecchi, 1991; Chong, 1991; Mahon, 1996). When only X_i or Y_i is subject to errors, the derivation of the method is straightforward. When both X_i and Y_i are subject to weighted errors, the mathematical treatment is complex, especially if the errors are correlated. A seemingly simple problem is not always simple, which is often true in both natural sciences and social sciences.

9.1. Least Squares: No Errors in X_i ; Y_i Subject to Equal Errors

When there is no error in X_i (Fig. 9.1), then $x_i = X_i$, and the sum of the squares of the residuals in Y_i become

$$S = \sum (y_i - Y_i)^2 = \sum (a + bx_i - Y_i)^2 = \sum (a + bX_i - Y_i)^2 . \quad (9.1)$$

To minimize S, the partial derivative of S related to a and b must be zero. From Eq. (9.1), we have,

$$\frac{\partial S}{\partial a} = 2\sum (a + bX_i - Y_i) = 0, \tag{9.2}$$

$$\frac{\partial S}{\partial b} = 2\sum X_i (a + bX_i - Y_i) = 0. \tag{9.3}$$

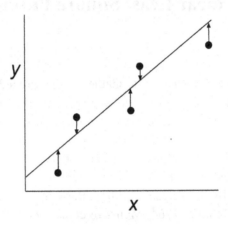

Fig. 9.1. Sketch showing the linear regression for the case when Y_i is subject to errors and there are no errors in X_i.

By rewriting (9.2) and (9.3), we obtain

$$aN + b\sum X_i - \sum Y_i = 0, \tag{9.4}$$

and

$$a\sum X_i + b\sum X_i^2 - \sum X_i Y_i = 0. \tag{9.5}$$

The solutions to the simultaneous equations (9.4) and (9.5) are

$$a = \frac{\sum X_i^2 \sum Y_i - \sum X_i \sum X_i Y_i}{\Delta}, \tag{9.6}$$

$$b = \frac{N\sum X_i Y_i - \sum X_i \sum Y_i}{\Delta}, \tag{9.7}$$

where

$$\Delta = N\sum X_i^2 - \left(\sum X_i\right)^2 . \tag{9.8}$$

The uncertainty in measurement y is given by a sum of squares with the following form:

$$\sigma_y = \sqrt{\frac{1}{N}\sum (Y_i - a - bX_i)^2} . \tag{9.9}$$

The uncertainties in a and b and are related to σ_y by

$$\sigma_a = \sigma_y \sqrt{\frac{\sum X_i^2}{\Delta}} , \tag{9.10}$$

$$\sigma_b = \sigma_y \sqrt{\frac{N}{\Delta}} . \tag{9.11}$$

Problem. Derive Eq. (9.10) for σ_a .

Differentiation of Eq. (9.6) with respect to Y_1, Y_2, ... Y_i, ... Y_N, respectively, gives

$$\frac{\partial a}{\partial Y_1} = \frac{\sum X_i^2 - X_1\left(\sum X_i\right)}{\Delta} , \tag{9.12}$$

$$\frac{\partial a}{\partial Y_i} = \frac{\sum X_i^2 - X_i\left(\sum X_i\right)}{\Delta} , \tag{9.13}$$

$$\frac{\partial a}{\partial Y_N} = \frac{\sum X_i^2 - X_N\left(\sum X_i\right)}{\Delta} . \tag{9.14}$$

Note that X_i is not subject to errors.

Using the error propagation principle, we have

$$\sigma_a = \sqrt{\left(\frac{\partial a}{\partial Y_1}\sigma_{Y_1}\right)^2 + ... + \left(\frac{\partial a}{\partial Y_i}\sigma_{Y_i}\right)^2 + ... + \left(\frac{\partial a}{\partial Y_N}\sigma_{Y_N}\right)^2} . \tag{9.15}$$

Since the uncertainties in Y are the same in this case, we have

$$\sigma_{Y_1} = \sigma_{Y_2} = ... \sigma_{Y_i} = ... = \sigma_{Y_n} = \sigma_Y . \tag{9.16}$$

Substitution of Eqs. (9.12), (9.13), (9.14) and (9.16) into (9.15) yields

$$\sigma_a = \sigma_Y \sqrt{\left(\frac{\partial a}{\partial Y_1}\right)^2 + ... \left(\frac{\partial a}{\partial Y_i}\right)^2 + ... + \left(\frac{\partial a}{\partial Y_n}\right)^2}$$

$$= \sigma_Y \sqrt{\left(\frac{\left[\sum X_i^2 - X_1\left(\sum X_i\right)\right]}{\Delta}\right)^2 + ... \left(\frac{\left[\sum X_i^2 - X_i\left(\sum X_i\right)\right]}{\Delta}\right)^2 + ... + \left(\frac{\left[\sum X_i^2 - X_n\left(\sum X_i\right)\right]}{\Delta}\right)^2}$$

$$= \frac{\sigma_Y}{\Delta} \sqrt{N\left(\sum X_i^2\right)^2 - 2\left(X_1 + ... X_i + ... + X_N\right)\left(\sum X_i^2\right)\left(\sum X_i\right) + \left(X_1^2 + ... + X_i^2 + ... + X_N^2\right)\left(\sum X_i\right)^2}$$

$$= \frac{\sigma_Y}{\Delta} \sqrt{\left(\sum X_i^2\right)\left[N\sum X_i^2 - \left(\sum X_i\right)^2\right]} = \sigma_Y \sqrt{\frac{\sum X_i^2}{\Delta}}.$$

$$(9.17)$$

Problem. Derive Eq. (9.11) for σ_b .

Differentiation of Eq. (9.7) with respect to Y_1, Y_2, ... Y_i ... Y_N, respectively, yields

$$\frac{\partial b}{\partial Y_1} = \frac{X_1 N - \sum X_i}{\Delta}, \qquad (9.18)$$

$$\frac{\partial b}{\partial Y_2} = \frac{X_2 N - \sum X_i}{\Delta}, \qquad (9.19)$$

$$\frac{\partial b}{\partial Y_i} = \frac{X_i N - \sum X_i}{\Delta}, \qquad (9.20)$$

$$\frac{\partial b}{\partial Y_N} = \frac{X_N N - \sum X_i}{\Delta}. \qquad (9.21)$$

From the principle of error propagation, the error in b is given by

$$\sigma_b = \sqrt{\left(\frac{\partial b}{\partial Y_1}\sigma_{Y_1}\right)^2 + ... + \left(\frac{\partial b}{\partial Y_i}\sigma_{Y_i}\right)^2 + ... + \left(\frac{\partial b}{\partial Y_N}\sigma_{Y_N}\right)^2}. \quad (9.22)$$

Assuming $\sigma_{Y_1} = ... = \sigma_{Y_i} = \sigma_{Y_N}$, and by substituting (9.18), (9.19), (9.20) and (9.21) into Eq. (9.22), we obtain

$$\sigma_b = \sigma_Y \sqrt{\left(\frac{\partial b}{\partial Y_1}\right)^2 + ... \left(\frac{\partial b}{\partial Y_i}\right)^2 + ... + \left(\frac{\partial b}{\partial Y_n}\right)^2}$$

$$= \sigma_y \sqrt{\left(\frac{X_1 N - \sum X_i}{\Delta}\right)^2 + ... \left(\frac{X_i N - \sum X_i}{\Delta}\right)^2 + ... + \left(\frac{X_N N - \sum X_i}{\Delta}\right)^2}$$

$$= \frac{\sigma_y}{\Delta} \sqrt{N^2\left(X_1^2 + ... X_i^2 + ... + X_N^2\right) - 2N\left(X_1 + ... X_i + ... + X_N\right)\sum X_i + N\left(\sum X_i\right)^2}$$

$$= \frac{\sigma_y}{\Delta} \sqrt{N\left[N\sum X_i^2 - \left(\sum X_i\right)^2\right]} = \sigma_y \sqrt{\frac{N}{\Delta}}.$$

$$(9.23)$$

Example. Assuming there is no error in X_i, calculate the slope and intercept of the straight line defined from the data in Table 9.1.

Table 9.1. (X_i , Y_i) data for linear regression.

	X_i	Y_i	X_i^2	$X_i Y_i$
1	0.2	5.2	0.04	1.04
2	2.9	7.4	8.41	21.46
3	5.1	8	26.01	40.80
4	7.1	10.9	50.41	77.39
5	10.0	13.7	100.00	137.00
6	13.9	16.5	193.21	229.35
7	19.3	21.3	372.49	411.09
N = 7	$\sum X_i$ = 58.5	$\sum Y_i$ = 83.0	$\sum X_i^2$ = 750.57	$\sum X_i Y_i$ = 918.13

The related parameters ($\sum X_i$, $\sum Y_i$, $\sum X_i^2$ and $\sum X_iY_i$) can be calculated and presented in Table 9.1.

From Eqs. (9.6), (9.7) and (9.8) we obtain

$$\Delta = N\sum X_i^2 - \left(\sum X_i\right)^2 = 7 \times 750.57 - 58.5^2 = 1831.74,$$

$$a = \frac{\sum X_i^2 \sum Y_i - \sum X_i \sum X_iY_i}{\Delta}$$

$$= \frac{750.57 \times 83.0 - 58.5 \times 918.13}{1831.74} = 4.688,$$

$$b = \frac{N\sum X_iY_i - \sum X_i \sum Y_i}{\Delta}$$

$$= \frac{7 \times 918.13 - 58.5 \times 83.0}{1831.74} = 0.858.$$

Thus the best fitting line is

$$y = 4.688 + 0.858x.$$

The linear regression line to the data in Table 9.1 is given in Fig. 9.2.

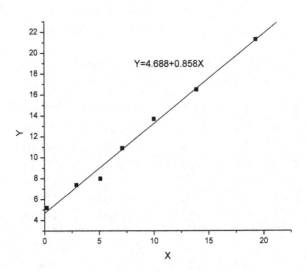

Fig. 9.2. Plot for the linear regression of the data from Table 9.1.

The sum of the squares of the residues can be obtained by

$$\sum (Y_i - a - bX_i)^2$$

$$= (5.2 - 4.688 - 0.858 \times 0.2)^2 + (7.4 - 4.688 - 0.858 \times 2.9)^2$$

$$+ (8.0 - 4.688 - 0.858 \times 5.1)^2 + (10.9 - 4.688 - 0.858 \times 7.1)^2$$

$$+ (13.7 - 4.688 - 0.858 \times 0.2)^2 + (16.5 - 4.688 - 0.858 \times 13.9)^2$$

$$+ (21.3 - 4.688 - 0.858 \times 19.3)^2 = 1.5145.$$

From Eq. (9.9) the uncertainty in y is

$$\sigma_Y = \sqrt{\frac{1}{N} \sum (Y_i - a - bX_i)^2} = \sqrt{\frac{1}{7} \times 1.5145} = 0.465 .$$

From Eq. (9.10) the uncertainty in the intercept a is

$$\sigma_a = \sigma_Y \sqrt{\frac{\sum X_i^2}{\Delta}} = 0.465 \times \sqrt{\frac{750.57}{1831.74}} = 0.298 .$$

From Eq. (9.11) the error in the slope b is

$$\sigma_b = \sigma_Y \sqrt{\frac{N}{\Delta}} = 0.465 \times \sqrt{\frac{7}{1831.74}} = 0.0288 .$$

Alternatively, a situation may arise when there are no errors in Y but X is subject to equal errors (Fig. 9.3).

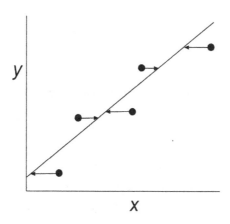

Fig. 9.3. Sketch showing the case where only X is subject to error.

For the situation in Fig. 9.3, by switching X_i with Y_i in all the equations (9.6), (9.7), (9.9), (9.10) and (9.11) for the case in Fig. 9.1 where only Y is subject to equal errors, we can obtain the corresponding equations for the case where only X is subject to equal errors (Fig. 4).

9.2. Least Squares: No Errors in X_i ; Y_i Subject to Weighted Errors

In session 9.1, we deal with the situation where Y_i is subject to equal errors. In this session, we investigate the case where Y_i is subject to different errors (Fig. 9.4).

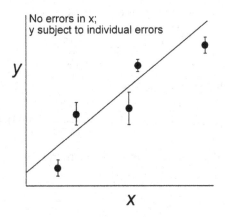

Fig. 9.4. Sketch showing that Y is subject to individual errors.

The sum of the squares of the residues with weighted errors are given by

$$S = \sum \omega_{Y_i} (y_i - Y_i)^2 = \sum \omega_{Y_i} (a + bx_i - Y_i)^2 . \qquad (9.24)$$

Note that when $\omega_{Y_i} = 1$ for all Y_i errors, Eq. (9.24) reduces to Eq. (9.1) for equal Y_i errors. To minimize S for best fitting, we need

$$\frac{\partial S}{\partial a} = 0 \text{ and } \frac{\partial S}{\partial b} = 0 .$$

Differentiation of Eq. (9.24) with respect to a and b, respectively, gives

$$\frac{\partial S}{\partial a} = 2\sum \omega_{Y_i}\left(a + bX_i - Y_i\right) = 0,$$ (9.25)

$$\frac{\partial S}{\partial b} = 2\sum \omega_{Y_i} X_i\left(a + bX_i - Y_i\right) = 0.$$ (9.26)

From (9.25) we get

$$a\sum \omega_{Y_i} + b\sum \omega_{Y_i} X_i + \sum \omega_{Y_i} Y_i = 0.$$ (9.27)

From (9.26) we have

$$a\sum \omega_{Y_i} X_i + b\sum \omega_{Y_i} X_i^2 - \sum \omega_{Y_i} X_i Y_i = 0.$$ (9.28)

The solutions to Eqs. (9.27) and (9.28) are

$$a = \frac{\sum \omega_{Y_i} X_i^2 \sum \omega_{Y_i} Y_i - \sum \omega_{Y_i} X_i \sum \omega_{Y_i} X_i Y_i}{\Delta},$$ (9.29)

$$b = \frac{\sum \omega_{Y_i} \sum \omega_{Y_i} X_i Y_i - \sum \omega_{Y_i} X_i \sum \omega_{Y_i} Y_i}{\Delta},$$ (9.30)

where

$$\Delta = \sum \omega_{Y_i} \sum \omega_{Y_i} X_i^2 - \left(\sum \omega_{Y_i} X_i\right)^2.$$ (9.31)

Problem. Derive the equation for the error in intercept, σ_a, when Y_i is subjected to weighted errors.
The partial derivatives of a with respect to Y_1, Y_2, Y_i, Y_N are

$$\frac{\partial a}{\partial Y_1} = \frac{\omega_{Y_1}}{\Delta}\left(\sum \omega_{Y_i} X_i^2 - X_1\sum \omega_{Y_i} X_i\right),$$ (9.32)

$$\frac{\partial a}{\partial Y_2} = \frac{\omega_{Y_2}}{\Delta}\left(\sum \omega_{Y_i} X_i^2 - X_2\sum \omega_{Y_i} X_i\right),$$ (9.33)

$$\frac{\partial a}{\partial Y_i} = \frac{\omega_{Y_i}}{\Delta}\left(\sum \omega_{Y_i} X_i^2 - X_i\sum \omega_{Y_i} X_i\right),$$ (9.34)

$$\frac{\partial a}{\partial Y_N} = \frac{\omega_{Y_N}}{\Delta}\left(\sum \omega_{Y_i} X_i^2 - X_N \sum \omega_{Y_i} X_i\right). \tag{9.35}$$

The uncertainty in a is therefore

$$\sigma_a = \sqrt{\left(\frac{\partial a}{\partial Y_1}\sigma_{Y_1}\right)^2 + ... + \left(\frac{\partial a}{\partial Y_i}\sigma_{Y_i}\right)^2 + ... + \left(\frac{\partial a}{\partial Y_N}\sigma_{Y_N}\right)^2}$$

$$= \sqrt{\begin{array}{l}\left(\dfrac{\omega_{Y_1}\left[\sum \omega_i X_i^2 - X_1\left(\sum \omega_i X_i\right)\right]}{\Delta}\right)^2 \sigma_{Y_1}^2 + ... \\[4mm] + \left(\dfrac{\omega_{Y_i}\left[\sum \omega_i X_i^2 - X_2\left(\sum \omega_i X_i\right)\right]}{\Delta}\right)^2 \sigma_{Y_i}^2 + ...\end{array}} \tag{9.36}$$

Since $\omega_{Y_1} = 1/\sigma_{Y_1}^2$, $\omega_{Y_2} = 1/\sigma_{Y_2}^2$, $\omega_{Y_i} = 1/\sigma_{Y_i}^2$, $\omega_{Y_N} = 1/\sigma_{Y_N}^2$, $\tag{9.37}$

substitution of (9.37) into Eq. (9.36) yields

$$\sigma_a = \sqrt{\frac{\sum \omega_i X_i^2}{\Delta}}. \tag{9.38}$$

Problem. Derive the equation for σ_b, the error in slope, when Y_i is subjected to weighted errors.

The partial derivatives of b with respect to Y_1, Y_2, Y_i, Y_N are

$$\frac{\partial b}{\partial Y_1} = \frac{\omega_{Y_1}}{\Delta}\left(X_1 \sum \omega_{Y_i} - \sum \omega_{Y_i} X_i\right), \tag{9.39}$$

$$\frac{\partial b}{\partial Y_2} = \frac{\omega_{Y_2}}{\Delta}\left(X_2 \sum \omega_{Y_i} - \sum \omega_{Y_i} X_i\right), \tag{9.40}$$

$$\frac{\partial b}{\partial Y_i} = \frac{\omega_{Y_i}}{\Delta}\left(X_i \sum \omega_{Y_i} - \sum \omega_{Y_i} X_i\right), \tag{9.41}$$

$$\frac{\partial b}{\partial Y_N} = \frac{\omega_{Y_N}}{\Delta}\left(X_N \sum \omega_{Y_i} - \sum \omega_{Y_i} X_i\right). \tag{9.42}$$

The uncertainty in b is therefore

$$\sigma_b = \sqrt{\left(\frac{\partial b}{\partial Y_1}\sigma_{Y_1}\right)^2 + \ldots \left(\frac{\partial b}{\partial Y_i}\sigma_{Y_i}\right)^2 + \ldots + \left(\frac{\partial b}{\partial Y_n}\sigma_{Y_N}\right)^2}$$

$$= \sqrt{\begin{array}{l} \dfrac{\omega_{Y_1}^2 \sigma_{Y_1}^2}{\Delta^2}\left(X_1\sum\omega_{Y_i} - \sum\omega_{Y_i}X_i\right)^2 + \ldots \\[2mm] + \dfrac{\omega_{Y_i}^2 \sigma_{Y_i}^2}{\Delta^2}\left(X_i\sum\omega_{Y_i} - \sum\omega_{Y_i}X_i\right)^2 + \ldots \\[2mm] + \dfrac{\omega_{Y_N}^2 \sigma_{Y_N}^2}{\Delta^2}\left(X_N\sum\omega_{Y_i} - \sum\omega_{Y_i}X_i\right)^2 \end{array}} \qquad (9.43)$$

Since $\omega_{Y_1} = 1/\sigma_{Y_1}^2$, $\omega_{Y_2} = 1/\sigma_{Y_2}^2$, $\omega_{Y_i} = 1/\sigma_{Y_i}^2$, $\omega_{Y_N} = 1/\sigma_{Y_N}^2$, we have

$$\sigma_b = \frac{1}{\Delta}\sqrt{\begin{array}{l}\left(\omega_{Y_1}X_1^2 + \omega_{Y_1}X_1^2 + \ldots + \omega_{Y_N}X_N^2\right)\left(\sum\omega_{Y_i}\right)^2 \\[2mm] -2\left(\omega_{Y_1}X_1 + \omega_{Y_1}X_1 + \ldots + \omega_{Y_N}X_N\right)\sum\omega_{Y_i}\sum\omega_{Y_i}X_i \\[2mm] +\left(\omega_{Y_1}X_1^2 + \omega_{Y_1}X_1^2 + \ldots + \omega_{Y_N}X_N^2\right)\left(\sum\omega_{Y_i}X_i\right)^2\end{array}} = \sqrt{\frac{\sum\omega_{Y_i}}{\Delta}}.$$

$$(9.44)$$

Example. Calculate the slope and intercept from the data in Table 9.2.

Table 9.2. Dataset (X_i, Y_i) and weight (ω_{Y_i}) for linear regression.

	X_i	Y_i	ω_{Y_i}	$\omega_{Y_i}X_i^2$	$\omega_{Y_i}X_i$	$\omega_{Y_i}Y_i$	$\omega_{Y_i}X_iY_i$
1	0.2	5.2	4	0.16	0.8	20.8	4.16
2	2.9	7.4	3	25.23	8.7	22.2	64.38
3	5.1	8	2	52.02	10.2	16.0	81.60
4	7.1	10.9	7	352.87	49.7	76.3	541.73
5	10	13.7	2	200.00	20.0	27.4	274.00
6	13.9	16.5	3	579.63	41.7	49.5	688.05
7	19.3	21.3	9	3352.41	173.7	191.7	3699.81
$N = 7$	$\sum X_i = 58.5$	$\sum Y_i = 83.0$	$\sum \omega_{Y_i} = 30.0$	$\sum \omega_{Y_i}X_i^2 = 4562.32$	$\sum \omega_{Y_i}X_i = 304.8$	$\sum \omega_{Y_i}Y_i = 403.9$	$\sum \omega_{Y_i}X_iY_i = 5353.73$

The related parameters are calculated and presented in Table 9.2. We obtain

$$\Delta = \sum \omega_{Y_i} \sum \omega_{Y_i} X_i^2 - \left(\sum \omega_{Y_i} X_i \right)^2$$
$$= 30.0 \times 4562.32 - 304.8^2 = 43966.56.$$

The intercept can be obtained from

$$a = \frac{4562.32 \times 403.9 - 304.8 \times 5353.73}{43966.56} = 4.797 .$$

Finally, the slope is given from

$$b = \frac{30.0 \times 5353.73 - 304.8 \times 403.9}{43966.56} = 0.853 .$$

Thus, the best fitting line when the weight is considered is

$$y = 4.797 + 0.853x .$$

Example. Calculate the slope and intercept from the following (X_i, Y_i) data when $\omega(Y_i)$ is calculated from $\omega_{Y_i} = 1/\sigma_{Y_i}^2$.

Table 9.3. (X_i, Y_i) data and weight (ω_{Y_i}) for linear regression.

	X_i	Y_i	σ_{Y_i}	ω_{Y_i}	$\omega_{Y_i} X_i^2$	$\omega_{Y_i} X_i$	$\omega_{Y_i} Y_i$	$\omega_{Y_i} X_i Y_i$
1	0.2	5.2	0.10	100.0	4.0	20.0	520.0	104.0
2	2.9	7.4	0.17	34.6	291.0	100.3	256.1	742.6
3	5.1	8	0.16	39.1	1016.0	199.2	312.5	1593.8
4	7.1	10.9	0.09	123.5	6223.5	876.5	1345.7	9554.3
5	10	13.7	0.13	59.2	5917.2	591.7	810.7	8106.5
6	13.9	16.5	0.16	39.1	7547.3	543.0	644.5	8959.0
7	19.3	21.3	0.08	156.3	58201.6	3015.6	3328.1	64232.8
N $= 7$	$\sum X_i$ $= 58.5$	$\sum Y_i$ $= 83.0$		$\sum \omega_{Y_i}$ $= 551.6$				

According to Table 9.3, we obtain

$$\sum \omega_{Y_i} X_i^2 = 79200.5 , \quad \sum \omega_{Y_i} X_i = 5346.4 ,$$

$$\sum \omega_{Y_i} Y_i = 7217.5, \quad \sum \omega_{Y_i} X_i Y_i = 93292.9,$$

$$\Delta = \sum \omega_{Y_i} \sum \omega_{Y_i} X_i^2 - \left(\sum \omega_{Y_i} X_i \right)^2$$

$$= 551.6 \times 79200.5 - 5346.4^2 = 15,103,226.3.$$

From (9.29) we get

$$a = \frac{79200.5 \times 7217.5 - 5346.4 \times 93292.9}{15103226.3} = 4.823.$$

From (9.30) we obtain

$$b = \frac{551.6 \times 93292.9 - 5346.4 \times 7217.5}{15103226.3} = 0.852.$$

From (9.38) we have

$$\sigma_a = \sqrt{\frac{\sum \omega_{Y_i} X_i^2}{\Delta}} = \sqrt{\frac{79200.5}{15103226.3}} = 0.072.$$

From (9.44) we get

$$\sigma_b = \sqrt{\frac{\sum \omega_{Y_i}}{\Delta}} = \sqrt{\frac{551.6}{15103226.3}} = 0.006.$$

The resulting linear regression line is plotted in Fig. 9.6.

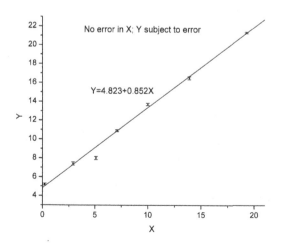

Fig. 9.5. Plot of linear regression of the data in Table 9.3.

9.3. Both X_i and Y_i are Subject to Weighted Errors

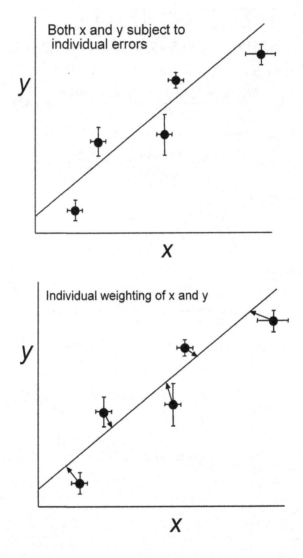

Fig. 9.6. Diagrams showing the situation when both X and Y are subject to errors and individual weighting of X and Y.

When both X_i and Y_i are subject to weighted errors (Fig. 9.6), the goal is to minimize the following expression:

$$S = \sum \left[\omega_{X_i} (x_i - X_i)^2 + \omega_{Y_i} (y_i - Y_i)^2 \right]$$
$$= \sum \left[\omega_{X_i} (x_i - X_i)^2 + \omega_{Y_i} (a + bx_i - Y_i)^2 \right]. \tag{9.45}$$

The conditions to minimize S include all the following three partial derivative of S to be zero:

$$\frac{\partial S}{\partial x_i} = 0, \tag{9.46}$$

$$\frac{\partial S}{\partial a} = 0, \tag{9.47}$$

$$\frac{\partial S}{\partial b} = 0. \tag{9.48}$$

Since $\dfrac{\partial S}{\partial x_i} = 0$, we have

$$\frac{\partial S}{\partial x_i} = 2\omega_{X_i} (x_i - X_i) + 2b\omega_{Y_i} (a + bx_i - Y_i)$$
$$= 2 \left[(x_i - X_i)\omega_{X_i} + b\omega_{Y_i} (a + bx_i - Y_i) \right] \tag{9.49}$$
$$= 0.$$

The solution of Eq. (9.49) is

$$x_i = \frac{X_i \omega_{X_i} + b\omega_{Y_i} (Y_i - a)}{\omega_{X_i} + b^2 \omega_{Y_i}}. \tag{9.50}$$

Substitution of Eq. (9.50) into Eq. (9.45) yields

$$S = \sum W_i (a + bX_i - Y_i)^2, \tag{9.51}$$

where

$$W_i = \frac{\omega_{X_i} \omega_{Y_i}}{\omega_{X_i} + b^2 \omega_{Y_i}} . \tag{9.52}$$

Since $\dfrac{\partial S}{\partial a} = 0$ (Eq. (9.47)), differentiating S with respect to a from Eq. (9.51), we obtain

$$\frac{\partial S}{\partial a} = \sum 2W_i \left(a + bX_i - Y_i \right) = 0 , \tag{9.53}$$

or

$$a \sum W_i + b \sum W_i X_i - \sum W_i Y_i = 0 , \tag{9.54}$$

$$a + b \frac{\sum W_i X_i}{\sum W_i} = \frac{\sum W_i Y_i}{\sum W_i} . \tag{9.55}$$

Therefore we have

$$a + b\overline{X} = \overline{Y} , \tag{9.56}$$

where

$$\overline{X} = \frac{\sum W_i X_i}{\sum W_i} , \tag{9.57}$$

and

$$\overline{Y} = \frac{\sum W_i Y_i}{\sum W_i} , \tag{9.58}$$

are weighted averages of X_i and Y_i, respectively. Substitution of Eq. (9.56) into Eq. (9.51) gives

$$S = \sum W_i \left(bU_i - V_i \right)^2 , \tag{9.59}$$

where

$$U_i = X_i - \overline{X} , \tag{9.60}$$

and

$$V_i = Y_i - \overline{Y} . \tag{9.61}$$

For the last partial derivative, $\dfrac{\partial S}{\partial b} = 0$ (Eq. (9.48)), differentiating S with respect to b in Eq. (9.59) gives rise to

$$\frac{\partial S}{\partial b} = \frac{\partial \sum W_i \left(bU_i - V_i \right)^2}{\partial b} = 0. \tag{9.62}$$

Note that W_i is also a function of b. From Eq. (9.52), we have

$$\frac{\partial W_i}{\partial b} = -2b\omega_{X_i}\omega_{Y_i}^{\;2}\left[\omega_{X_i} + b^2\omega_{Y_i} \right]^{-2} = -\frac{2bW_i^2}{\omega_{X_i}}. \tag{9.63}$$

From Eq. (9.62) we obtain

$$\frac{\partial S}{\partial b} = \frac{\partial \sum W_i \left(bU_i - V_i \right)^2}{\partial b}$$
$$= \sum \left[2W_i \left(bU_i - V_i \right)U_i + \frac{\partial W_i}{\partial b}\left(bU_i - V_i \right)^2 \right]. \tag{9.64}$$

Combination of Eqs. (9.63) and (9.64) gives rise to

$$\frac{\partial S}{\partial b} = \sum \left[2W_iU_i \left(bU_i - V_i \right) - 2\frac{bW_i^2}{\omega_{X_i}}\left(bU_i - V \right)^2 \right] = 0. \tag{9.65}$$

that is

$$b^3 \sum \frac{W_i^2 U_i^2}{\omega_{X_i}} - 2b^2 \sum \frac{W_i^2 U_i V_i}{\omega_{X_i}} - b\left(\sum W_iU_i^2 - \sum \frac{W_i^2 V_i^2}{\omega_{X_i}} \right)$$
$$+ \sum W_iU_iV_i = 0. \tag{9.66}$$

This is the York's "Least Square Cubic" (York, 1966). Equation (9.66) can be simplified to get rid of b^3. The key is to make good use of W_i. Equation (9.66) can be written as

$$b^3 \sum \frac{W_i^2 U_i^2}{\omega_{X_i}} - 2b^2 \sum \frac{W_i^2 U_i V_i}{\omega_{X_i}} + b \sum \frac{W_i^2 V_i^2}{\omega_{X_i}}$$
$$= b \sum W_iU_i^2 - \sum W_iU_iV_i. \tag{9.67}$$

Since $W_i = \dfrac{\omega_{X_i}\omega_{Y_i}}{\omega_{X_i} + b^2\omega_{Y_i}}$, we have

$$W_i\left(v_i + b^2 u_i\right) = 1,\tag{9.68}$$

or

$$\frac{W_i\left[\omega_{X_i} + b^2\omega_{Y_i}\right]}{\omega_{X_i}\omega_{Y_i}} = 1.\tag{9.69}$$

Multiple by 1, or $\dfrac{W_i\left[\omega_{X_i} + b^2\omega_{Y_i}\right]}{\omega_{X_i}\omega_{Y_i}}$ according to the definition of W_i, to

the right hand side of Eq. (9.67), the right hand side becomes

$$b\sum W_i U_i^2 - \sum W_i U_i V_i$$

$$= b\sum W_i X_i'^2 \frac{W_i\left[\omega_{X_i} + b^2\omega_{Y_i}\right]}{\omega_{X_i}\omega_{Y_i}} - \sum W_i X_i' Y_i' \frac{W_i\left[\omega_{X_i} + b^2\omega_{Y_i}\right]}{\omega_{X_i}\omega_{Y_i}}\tag{9.70}$$

$$= b^3\sum \frac{W_i^2 U_i^2}{\omega_{X_i}} - b^2\sum \frac{W_i^2 U_i V_i}{\omega_{X_i}} + b\sum \frac{W_i^2 U_i}{\omega_{Y_i}} - \sum \frac{W_i^2 U_i V_i}{\omega_{Y_i}}.$$

Combination of Eq. (9.67) and (9.70) gives

$$b^3\sum \frac{W_i^2 U_i^2}{\omega_{X_i}} - 2b^2\sum \frac{W_i^2 U_i V_i}{\omega_{X_i}} + b\sum \frac{W_i^2 V_i^2}{\omega_{X_i}}$$

$$= b^3\sum \frac{W_i^2 U_i^2}{\omega_{X_i}} - b^2\sum \frac{W_i^2 U_i V_i}{\omega_{X_i}} + b\sum \frac{W_i^2 U_i}{\omega_{Y_i}} - \sum \frac{W_i^2 U_i V_i}{\omega_{Y_i}},\tag{9.71}$$

or

$$b^2\sum W_i^2 U_i V_i/\omega_{X_i} + b\left(\sum W_i^2 U_i^2/\omega_{Y_i} - \sum W_i^2 V_i^2/\omega_{X_i}\right)$$

$$- \sum W_i^2 U_i V_i/\omega_{Y_i} = 0.\tag{9.72}$$

Equation (9.72) can still be further simplified. The key is to move b
inside $\sum$ in selected terms. Rewriting Eq. (9.72), we obtain

$$b^2 \sum W_i^2 U_i V_i \big/ \omega_{X_i} + b \sum W_i^2 U_i^2 \big/ \omega_{Y_i}$$
$$= b \sum W_i^2 V_i^2 \big/ \omega_{X_i} + \sum W_i^2 U_i V_i \big/ \omega_{Y_i}. \tag{9.73}$$

By placing b in the first and the third terms inside the sum, we have

$$b \sum b W_i^2 U_i V_i \big/ \omega_{X_i} + b \sum W_i^2 U_i^2 \big/ \omega_{Y_i}$$
$$= \sum b W_i^2 V_i^2 \big/ \omega_{X_i} + \sum W_i^2 U_i V_i \big/ \omega_{Y_i}, \tag{9.74}$$

or

$$b = \frac{\sum b W_i^2 V_i^2 \big/ \omega_{X_i} + \sum W_i^2 U_i V_i \big/ \omega_{Y_i}}{\sum b W_i^2 U_i V_i \big/ \omega_{X_i} + \sum W_i^2 U_i^2 \big/ \omega_{Y_i}}$$
$$= \frac{\sum W_i^2 V_i \left[U_i \big/ \omega_{Y_i} + b V \big/ \omega_{X_i} \right]}{\sum W_i^2 V_i \left[U_i \big/ \omega_{Y_i} + b V \big/ \omega_{X_i} \right]}. \tag{9.75}$$

Equation (9.75) can be written as

$$b = \frac{\sum W_i V_i Z_i}{\sum W_i U_i Z_i}, \tag{9.76}$$

where

$$Z_i = W_i \left(\frac{U_i}{\omega_{Y_i}} + \frac{b V_i}{\omega_{X_i}} \right). \tag{9.77}$$

Equations (9.76) and (9.77) are the identical results from Williamson, (1968).

To check if Eq. (9.76) collapses to least-square fitting for the case when Y_i is subject to equal errors and there are no errors in X_i ($\omega_{X_i} = \infty$, $\omega_{Y_i} = $ constant), we have

$$W_i = \text{constant}, \ Z_i = U_i$$

then Eq. (9.76) reduces to

$$b = \frac{\sum U_i V_i}{\sum U_i U_i} = \frac{\sum \left(X_i - \overline{X} \right)\left(Y_i - \overline{Y} \right)}{\sum \left(X_i - \overline{X} \right)^2} = \frac{N \sum X_i Y_i - \sum X_i \sum Y_i}{\Delta},$$

$$\tag{9.78}$$

which is Eq. (9.7).

Two special cases for errors in both X_i and Y_i are given below.

1) Equal weighing of X_i and Y_i (Fig. 9.8), i.e., $\omega_{X_i} = \omega_{Y_i}$, then

$$W_i = \frac{\omega_{X_i}}{b^2 + 1}.$$

From Eq. (9.76) we get

$$b = \frac{\sum (U_i + bV_i)V_i}{\sum (U_i + bV_i)U_i},$$

or

$$b^2 \sum U_i V_i + b\left(\sum U_i^2 - \sum V_i^2\right) - \sum U_i V_i = 0,$$

with the following solution:

$$b = \frac{\sum V_i^2 - \sum U_i^2 + \sqrt{\left(\sum U_i^2 - \sum V_i^2\right)^2 - 4\left(\sum U_i V_i\right)^2}}{2\sum U_i V_i}. \quad (9.79)$$

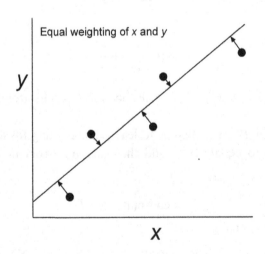

Equal weighting of x and y

Fig. 9.7. Least square fitting for equal weighing of X_i and Y_i. The lines of adjustment are perpendicular to the best fitting line.

2) The ratio of the weights is a constant, $\omega_{X_i} / \omega_{Y_i} = c$ (a constant)

$$W_i = \frac{\omega_{X_i}}{b^2 + c} .$$

From Eq. (9.76) we obtain

$$b = \frac{\sum \omega_{X_i} (cU_i + bV_i)V_i}{\sum \omega_{X_i} (cU_i + bV_i)U_i} ,$$

or

$$b^2 \sum \omega_{X_i} U_i V_i + b\left(c\sum \omega_{X_i} U_i^2 - \sum \omega_{X_i} V_i^2\right) - c\sum \omega_{X_i} U_i V_i = 0 .$$

Therefore, we have (Worthing and Geffner, 1946)

$$b = \frac{\sum V_i^2 - c\sum U_i^2 + \sqrt{\theta}}{2\sum U_i V_i} , \tag{9.80}$$

where

$$\theta = \left(c\sum U_i^2 - \sum V_i^2\right)^2 - 4\left(\sum U_i V_i\right)^2 . \tag{9.81}$$

9.4. Both X_i and Y_i are Subject to Weighted Correlated Errors

9.4.1. Formulation

When the weighted errors for X_i and Y_i are correlated, in order to achieve the goal of best linear fitting, we need to minimize the following expression (York, 1969):

$$S = \sum \frac{\omega_{X_i}(x_i - X_i)^2 - 2r_i\sqrt{\omega_{X_i}\omega_{Y_i}}(x_i - X_i)(y_i - Y_i) + \omega_{Y_i}(y_i - Y_i)^2}{1 - r_i^2}$$

$$= \sum \frac{\omega_{X_i}(x_i - X_i)^2 - 2r_i\sqrt{\omega_{X_i}\omega_{Y_i}}(x_i - X_i)(a + bx_i - Y_i) + \omega_{Y_i}(a + bx_i - Y_i)^2}{1 - r_i^2} .$$

$$\tag{9.82}$$

Again, we need to have the following three partial derivative of S to be zero:

$$\frac{\partial S}{\partial x_i} = 0, \tag{9.83}$$

$$\frac{\partial S}{\partial a} = 0, \tag{9.84}$$

$$\frac{\partial S}{\partial b} = 0. \tag{9.85}$$

According to Eq. (9.83), $\dfrac{\partial S}{\partial x_i} = 0$, we have

$$\frac{\partial S}{\partial x_i} = 2\omega_{X_i}(x_i - X_i) - 2r_i\sqrt{\omega_{X_i}\omega_{Y_i}}\left[(a + bx_i - Y_i) + b(x_i - X_i)\right]$$
$$+2b\omega_{Y_i}(a + bx_i - Y_i) \tag{9.86}$$
$$= 0.$$

The solution of Eq. (9.86) for x_i is

$$x_i = \frac{X_i\omega_{X_i} + r_i(a - bX_i - Y_i)\sqrt{\omega_{X_i}\omega_{Y_i}} - b\omega_{Y_i}(a - Y_i)}{b^2\omega_{Y_i} + \omega_{X_i} - 2br_i\sqrt{\omega_{X_i}\omega_{Y_i}}}. \tag{9.87}$$

Substitution of Eq. (9.87) into (9.82) gives

$$S = \sum W_i(a + bX_i - Y_i)^2, \tag{9.88}$$

where

$$W_i = \frac{\omega_{X_i}\omega_{Y_i}}{b^2\omega_{Y_i} + \omega_{X_i} - 2br_i\sqrt{\omega_{X_i}\omega_{Y_i}}}. \tag{9.89}$$

Note that when the errors are not correlated, that is, $r_i = 0$, Eq. (9.89) reduces to Eq. (9.52).

Since $\partial S/\partial a = 0$ (Eq. (9.84)), by differentiating S with respect to a from Eq.(9.88), we obtain

$$\frac{\partial S}{\partial a} = \sum 2W_i(a + bX_i - Y_i) = 0, \tag{9.90}$$

or

$$a\sum W_i + b\sum W_i X_i - \sum W_i Y_i = 0 , \quad (9.91)$$

$$a + b\frac{\sum W_i X_i}{\sum W_i} = \frac{\sum W_i Y_i}{\sum W_i} , \quad (9.92)$$

$$a + b\overline{X} = \overline{Y} , \quad (9.93)$$

where

$$\overline{X} = \frac{\sum W_i X_i}{\sum W_i} , \quad (9.94)$$

and

$$\overline{Y} = \frac{\sum W_i Y_i}{\sum W_i} , \quad (9.95)$$

are averages of X_i and Y_i.

Substitution of Eq. (9.93) into Eq. (9.88) gives

$$S = \sum W_i \left(bU_i - V_i \right)^2 , \quad (9.96)$$

where

$$U_i = X_i - \overline{X} , \quad (9.97)$$

and

$$V_i = Y_i - \overline{Y} , \quad (9.98)$$

and W_i is given by Eq. (9.89).

According to Eq. (9.85), differentiation of S with respect to b in Eq. (9.96) gives rise to

$$\frac{\partial S}{\partial b} = \frac{\partial \sum W_i \left(bU_i - V_i \right)^2}{\partial b} = 0 . \quad (9.99)$$

Note that W_i is also a function of b. Differentiation of W_i from Eq. (9.89) gives

$$\frac{\partial W_i}{\partial b} = -2b\omega_{X_i}\omega_{Y_i}^2 \left[\omega_{X_i} + b^2\omega_{Y_i} - 2br_i\sqrt{\omega_{X_i}\omega_{Y_i}} \right]^{-2}$$

$$= -2W_i^2 \left(\frac{b}{\omega_{X_i}} - \frac{r_i}{\sqrt{\omega_{X_i}\omega_{Y_i}}} \right). \tag{9.100}$$

From Eq. (9.99), we obtain

$$\frac{\partial S}{\partial b} = \frac{\partial \sum W_i (bU_i - V_i)^2}{\partial b}$$

$$= \sum \left[2W_i (bU_i - V_i) U_i + \frac{\partial W_i}{\partial b} (bU_i - V_i)^2 \right]. \tag{9.101}$$

Combination of Eqs. (9.100) and (9.101) leads to

$$\frac{\partial S}{\partial b} = \sum \left[2W_i U_i (bU_i - V_i) - 2 \left(\frac{bW_i^2}{\omega_{X_i}} - \frac{W_i^2 r_i}{\sqrt{\omega_{X_i}\omega_{Y_i}}} \right) (bU_i - V)^2 \right] = 0,$$

$$\tag{9.102}$$

or

$$b^3 \sum \frac{W_i^2 U_i^2}{\omega_{X_i}} - b^2 \left[2\sum \frac{W_i^2 U_i V_i}{\omega_{X_i}} + \sum \frac{W_i^2 r_i U_i^2}{\sqrt{\omega_{X_i}\omega_{Y_i}}} \right]$$

$$-b \left(\sum W_i U_i^2 - 2\sum \frac{W_i^2 r_i U_i V_i}{\sqrt{\omega_{X_i}\omega_{Y_i}}} - \sum \frac{W_i^2 V_i^2}{\omega_{X_i}} \right) \tag{9.103}$$

$$+ \sum W_i U_i V_i - \sum \frac{W_i^2 r_i V_i^2}{\sqrt{\omega_{X_i}\omega_{Y_i}}} = 0.$$

Equation (9.103) can be simplified to get rid of b^3 by making good use of W_i from Eq. (9.89). By using the similar techniques in section 9.3, Eq. (9.103) can be written as

$$b^2 \sum \left[\frac{W_i^2 U_i V_i}{\omega_{X_i}} - \frac{W_i^2 r_i U_i^2}{\sqrt{\omega_{X_i} \omega_{Y_i}}} \right] + b \left(\sum \frac{W_i^2 U_i^2}{\omega_{X_i}} - \sum \frac{W_i^2 V_i^2}{\omega_{X_i}} \right)$$

$$-\sum \left[\frac{W_i^2 U_i V_i}{\omega_{X_i}} - \frac{W_i^2 r_i V_i^2}{\sqrt{\omega_{X_i} \omega_{Y_i}}} \right] = 0,$$

(9.104)

Further simplification of Eq. (9.104) leads to

$$b = \frac{\sum W_i V_i Z_i}{\sum W_i U_i Z_i},$$

(9.105)

where

$$Z_i = \sum W_i \left[\frac{U_i}{\omega_{Y_i}} + \frac{b V_i}{\omega_{X_i}} - \frac{r_i V_i}{\sqrt{\omega_{X_i} \omega_{Y_i}}} \right].$$

(9.106)

9.4.2. Error analysis

Equation (9.104) can be written as

$$\vartheta = c_1 b^2 + c_2 b + c_3 = 0,$$

(9.107)

where

$$c_1 = \sum \left[\frac{W_i^2 U_i V_i}{\omega_{X_i}} - \frac{W_i^2 r_i U_i^2}{\sqrt{\omega_{X_i} \omega_{Y_i}}} \right],$$

$$c_2 = \sum \frac{W_i^2 U_i^2}{\omega_{X_i}} - \sum \frac{W_i^2 V_i^2}{\omega_{X_i}},$$

$$c_3 = -\sum \left[\frac{W_i^2 U_i V_i}{\omega_{X_i}} - \frac{W_i^2 r_i V_i^2}{\sqrt{\omega_{X_i} \omega_{Y_i}}} \right].$$

The error-propagation equation for the slope is (York, 1969)

$$\sigma_b^{\;2}=\left\{\sum_{i=1}^{n}\left[\left[\left(\frac{\partial\vartheta}{\partial x_i}\right)^2\sigma_{x_i}^{\;2}+\left(\frac{\partial\vartheta}{\partial y_i}\right)^2\sigma_{y_i}^{\;2}\right]\\+2r_i\sigma_{x_i}\sigma_{y_i}\left(\frac{\partial\vartheta}{\partial x_i}\right)\left(\frac{\partial\vartheta}{\partial y_i}\right)\right]\right\}\Bigg/\left(\frac{\partial\vartheta}{\partial b}\right)^2,\qquad(9.108)$$

and the error propagation for the intercept is

$$\sigma_a^{\;2}=\sum_{i=1}^{n}\left[\left(\frac{\partial a}{\partial x_i}\right)^2\sigma_{x_i}^{\;2}+\left(\frac{\partial a}{\partial y_i}\right)^2\sigma_{y_i}^{\;2}+2r_i\sigma_{x_i}\sigma_{y_i}\left(\frac{\partial a}{\partial x_i}\right)\left(\frac{\partial a}{\partial y_i}\right)\right].\qquad(9.109)$$

σ_{x_i} and σ_{y_i} are the errors for x_i and y_i, respectively.

Note that since

$$\left(\frac{\partial\vartheta}{\partial x_i}\right)^2\Bigg/\left(\frac{\partial\vartheta}{\partial b}\right)^2=\left(\frac{\partial b}{\partial x_i}\right)^2,$$

$$\left(\frac{\partial\vartheta}{\partial y_i}\right)^2\Bigg/\left(\frac{\partial\vartheta}{\partial b}\right)^2=\left(\frac{\partial b}{\partial y_i}\right)^2,$$

$$2r_i\sigma_{x_i}\sigma_{y_i}\left(\frac{\partial\vartheta}{\partial x_i}\right)\left(\frac{\partial\vartheta}{\partial y_i}\right)\Bigg/\left(\frac{\partial\vartheta}{\partial b}\right)^2=2r_i\sigma_{x_i}\sigma_{y_i}\left(\frac{\partial b}{\partial x_i}\right)\left(\frac{\partial b}{\partial y_i}\right),$$

Eq. (9.108) for σ_b has the identical expression as Eq. (9.109) for σ_a. Equation (9.108) is written in a manner so that we can use Eq. (9.107) to obtain partial derivatives.

The three partial derivatives in Eq. (9.108) are given as follows (Mahon, 1996):

$$\frac{\partial\vartheta}{\partial x_i}=\sum_{j=1}^{n}W_j^{\;2}\left[\delta_{ij}-\left(W_i\Bigg/\sum_{k=1}^{n}W_k\right)\right]\left[\begin{array}{c}b^2\left(V_j\sigma_{x_j}^{\;2}-2U_j r_j\sigma_j\sigma_j\right)\\+2bU_j\sigma_{yj}^{\;2}-V_j\sigma_{yj}^{\;2}\end{array}\right],\qquad(9.110)$$

$$\frac{\partial\vartheta}{\partial y_i}=\sum_{j=1}^{n}W_j^{\;2}\left[\delta_{ij}-\left(W_i\Bigg/\sum_{k=1}^{n}W_k\right)\right]\left[\begin{array}{c}b^2\left(U_j\sigma_{x_j}^{\;2}+2V_j r_j\sigma_{x_j}\sigma_{y_j}\right)\\-2bV_j\sigma_{xj}^{\;2}-U_j\sigma_{xj}^{\;2}\end{array}\right],\qquad(9.111)$$

$$\frac{\partial \vartheta}{\partial b} = \sum_{i=1}^{n} W_i^2 \left[2b\left(U_i V_i \sigma_{x_i}^2 - U_i^2 r_i \sigma_{x_i} \sigma_{y_i}\right) + \left(U_i^2 \sigma_{y_i}^2 - V_i^2 \sigma_{x_i}^2\right)\right]$$

$$+4\sum_{i=1}^{n} W_i^3 \left(r_i \sigma_{x_i} \sigma_{y_i} - b\sigma_{x_i}^2\right) \begin{bmatrix} b^2 \left(U_i V_i \sigma_{x_i}^2 - U_i^2 r_i \sigma_{x_i} \sigma_{y_i}\right) \\ +b\left(U_i^2 \sigma_{y_i}^2 - V_i^2 \sigma_{x_i}^2\right) \\ -\left(U_i V_i \sigma_{y_i}^2 - V_i^2 r_i \sigma_{x_i} \sigma_{y_i}\right) \end{bmatrix},$$

$$(9.112)$$

where δ_{ij} is the Kronecker delta and

$$\delta_{ij} = 0 \text{ for } i \neq j,$$

and

$$\delta_{ij} = 0 \text{ for } i = j \ (i \text{ and } j \text{ are integers}).$$

The two partial derivatives in Eq. (9.109) can be obtained by differentiating Eq. (9.93) (Mahon, 1996)

$$\frac{\partial a}{\partial x_i} = -\frac{bW_i}{\sum_{k=1}^{n} W_k} - \overline{X}\left(\frac{\partial \vartheta}{\partial x_i}\right) \bigg/ \left(\frac{\partial \vartheta}{\partial b}\right), \qquad (9.113)$$

$$\frac{\partial a}{\partial y_i} = \frac{W_i}{\sum_{k=1}^{n} W_k} - \overline{X}\left(\frac{\partial \vartheta}{\partial y_i}\right) \bigg/ \left(\frac{\partial \vartheta}{\partial b}\right). \qquad (9.114)$$

9.5. Summary of Linear Least Square Fitting of the Line $y = a + bx$

1) No errors in X_i; Y_i subject to equal errors

$$a = \frac{\sum X_i^2 \sum Y_i - \sum X_i \sum X_i Y_i}{\Delta}, \qquad b = \frac{N\sum X_i Y_i - \sum X_i \sum Y_i}{\Delta},$$

$$\text{where } \Delta = N\sum X_i^2 - \left(\sum X_i\right)^2.$$

The uncertainties in a and b are

$$\sigma_a = \sigma_y\sqrt{\frac{\sum X_i^2}{\Delta}}, \qquad\qquad \sigma_b = \sigma_y\sqrt{\frac{N}{\Delta}},$$

$$\text{where } \sigma_y = \sqrt{\frac{1}{N}\sum(Y_i - a - bX_i)^2}.$$

2) No errors in X_i; Y_i subject to weighted errors

$$a = \frac{\sum \omega_{Y_i} X_i^2 \sum \omega_{Y_i} Y_i - \sum \omega_{Y_i} X_i \sum \omega_{Y_i} X_i Y_i}{\Delta},$$

$$b = \frac{\sum \omega_{Y_i} \sum \omega_{Y_i} X_i Y_i - \sum \omega_{Y_i} X_i \sum \omega_{Y_i} Y_i}{\Delta},$$

where $\Delta = \sum \omega_{Y_i} \sum \omega_{Y_i} X_i^2 - \left(\sum \omega_{Y_i} X_i\right)^2$.

The uncertainties in a and b are

$$\sigma_a = \sqrt{\frac{\sum \omega_i X_i^2}{\Delta}}, \qquad\qquad \sigma_b = \sqrt{\frac{\sum \omega_{Y_i}}{\Delta}}.$$

3) Both X_i and Y_i are subject to weighted independent errors

$$a = \overline{Y} - b\overline{X}, \qquad\qquad b = \frac{\sum W_i V_i Z_i}{\sum W_i U_i Z_i},$$

where

$$\overline{X} = \frac{\sum W_i X_i}{\sum W_i}, \qquad\qquad \overline{Y} = \frac{\sum W_i Y_i}{\sum W_i},$$

$$W_i = \frac{\omega_{X_i} \omega_{Y_i}}{\omega_{X_i} + b^2 \omega_{Y_i}}, \qquad\qquad Z_i = W_i\left(\frac{U_i}{\omega_{Y_i}} + \frac{bV_i}{\omega_{X_i}}\right),$$

$$U_i = X_i - \overline{X}, \qquad\qquad V_i = Y_i - \overline{Y}.$$

4) Both X_i and Y_i are subject to weighted correlated errors

$$a = \overline{Y} - b\overline{X}, \qquad\qquad b = \frac{\sum W_i V_i Z_i}{\sum W_i U_i Z_i},$$

where

$$\overline{X} = \frac{\sum W_i X_i}{\sum W_i}, \qquad\qquad \overline{Y} = \frac{\sum W_i Y_i}{\sum W_i},$$

$$W_i = \frac{\omega_{X_i}\omega_{Y_i}}{b^2\omega_{Y_i} + \omega_{X_i} - 2br_i\sqrt{\omega_{X_i}\omega_{Y_i}}}, \qquad Z_i = \sum W_i \left[\frac{U_i}{\omega_{Y_i}} + \frac{bV_i}{\omega_{X_i}} - \frac{r_i V_i}{\sqrt{\omega_{X_i}\omega_{Y_i}}} \right],$$

$$U_i = X_i - \overline{X}, \qquad\qquad V_i = Y_i - \overline{Y}.$$

References

Adcock, R.J., 1878. A problem in least squares. The analyst, 5: 53-54.

Bartlett, M.S., 1949. Fitting a straight line when variables are subject to error. Bometrics, 5: 207-212.

Cecchi, G.C., 1991. Error analysis of the parameters of a least-squares determined curve when both variables have uncertainties. Meas. Sci. Technol., 2: 1127-1128.

Chong, D.P., 1991. Comment on linear leat-squares fits with errors in both variables. Am. J. Phys., 59: 472.

Deming, W.E., 1964. Statistical adjustment of data. Dovers Publications, New York.

Kermack, K.A. and Haldane, J.B.S., 1950. Organic correlation and allometry. Biometrica, 37: 30.

Lybanon, M., 1984. A better least-squares method when both variables have uncertainties. Am. J. Phys., 52: 22-26.

Mahon, K.I., 1996. The new "York" regression: Application of an improved statistical method to geochemistry. Int. Geol. Rev., 38: 293-303.

McIntyre, G.A., Brooks, C., Compston, W. and Turek, A., 1966. The statistical assessment of Rb-Sr isochrons. J. Geophys. Res., 71: 5459-5468.

Pearson, K., 1901. On lines and planes of closest fit to systems of points in space. Phil. Trans., 2: 559-572.

Reed, B.C., 1989. Linear least-squares fits with errors in both coordinates. Am. J. Phys., 57: 642-646.

Williamson, J.H., 1968. Least-squares fitting of a straight line. Canadian J. Physics, 46: 1845-1847.

Worthing, A.G. and Geffner, J., 1946. Treatment of experimental data. John Wiley and Sons, New York.

York, D., 1966. Least-squares fitting of a straight line. Canadian J. Physics, 44: 1079-1086.

York, D., 1967. The best isochron. Earth Planet. Sci. Lett., 2: 479-482.

York, D., 1969. Least squares fitting of a straight line with correlated errors. Earth Planet. Sci. Lett., 5: 320-324.

Chapter 10

Mass Fractionation in Ionization Processes

10.1. Mass Fractionation Laws

Mass spectrometers measure ions rather than neutral atoms. The ionization processes in mass spectrometers involve breaking of chemical bonds. The strength of these chemical bonds is mass dependent. Therefore, ionization processes lead to mass dependent fractionation. Because the potential energy well of the bond involving the lighter isotope is always shallower than that for heavier isotope, the bond with the lighter isotope is more readily broken, resulting in mass dependent fractionation. Thus, the measured isotopic ratios are not the true isotope ratios. For example, $^{206}Pb/^{204}Pb$ ratio measured by thermal ionization mass spectrometer or secondary mass spectrometer is smaller than its true value, because these machines preferentially ionize light isotope ^{204}Pb than heavy isotope ^{206}Pb. To improve the precision and accuracy of the measurement of isotopic compositions, the mass fractionation in mass spectrometers needs to be corrected.

If an element has two or more natural non-radiogenic isotopes, their isotopic ratio can be used to correct the mass fractionation in other isotope ratios of the same element. For example, $^{146}Nd/^{144}Nd$ (=0.7219) in natural samples can be used to correct mass fractionation for $^{143}Nd/^{144}Nd$, where ^{143}Nd is a radiogenic isotope from parent ^{147}Sm. As another example, $^{86}Sr/^{88}Sr$ (=0.1194) in natural samples is used to correct mass fractionation in mass spectrometer for $^{87}Sr/^{86}Sr$ ratio, where ^{87}Sr is a radiogenic isotope from parent ^{87}Rb.

There are three models for mass fractionation of isotopes in the ionization processes: a linear law, a power law and an exponential law (Wasserburg et al., 1981; Lee et al., 2001).

10.1.1. Linear law

Let R_{ij} be the isotopic ratio of isotopes i and j with masses m_i and m_j. The mass difference between the two isotopes is

$$\Delta m_{ij} = m_i - m_j .$$

The measured isotopic ratio is R_{ij}^M and the corrected isotopic ratio is R_{ij}^C. For the linear law, the corrected ratio is given by

$$R_{ij}^C = R_{ij}^M \left(1 + \delta \Delta m_{ij}\right), \tag{10.1}$$

where the fractionation factor δ is determined by a pair of non-radiogenic isotopes (u and v) of constant natural abundance

$$\delta = \left(\frac{R_{uv}^N}{R_{uv}^M} - 1\right) \Big/ \Delta m_{uv} , \tag{10.2}$$

and u and v represent a particular choice of an isotope pair for normalization, R_{uv}^M is the measured isotopic ratio, R_{ij}^N is reference (true) value and

$$\Delta m_{uv} = m_u - m_v .$$

Take Sr isotope as an example,

$$\begin{aligned}
\left(\frac{^{87}Sr}{^{86}Sr}\right)_{true} &= \left(\frac{^{87}Sr}{^{86}Sr}\right)_{measured} \left[1 + (87 - 86)\delta\right] \\
&= \left(\frac{^{87}Sr}{^{86}Sr}\right)_{measured} \left[1 + \delta\right],
\end{aligned} \tag{10.3}$$

where δ is calculated by

$$\delta = \left[\frac{\left({}^{86}Sr / {}^{88}Sr \right)_{true}}{\left({}^{86}Sr / {}^{88}Sr \right)_{measured}} - 1 \right] \bigg/ (86 - 88)$$

$$= \frac{1}{2} \left[1 - \frac{0.1194}{\left({}^{86}Sr / {}^{88}Sr \right)_{measured}} \right].$$

(10.4)

For Nd isotopes,

$$\left(\frac{{}^{143}Nd}{{}^{144}Nd} \right)_{true} = \left(\frac{{}^{143}Nd}{{}^{144}Nd} \right)_{measured} \left[1 + (143 - 144)\delta \right]$$

$$= \left(\frac{{}^{143}Nd}{{}^{144}Nd} \right)_{measured} \left[1 - \delta \right],$$

(10.5)

where

$$\delta = \left[\frac{\left({}^{146}Nd / {}^{144}Nd \right)_{true}}{\left({}^{146}Nd / {}^{144}Nd \right)_{measured}} - 1 \right] \bigg/ (146 - 144)$$

$$= \frac{1}{2} \left[\frac{0.7219}{\left({}^{146}Nd / {}^{144}Nd \right)_{measured}} - 1 \right].$$

(10.6)

10.1.2. Power law

For power law, the corrected ratio is provided by

$$R_{ij}^{C} = R_{ij}^{M} \left(1 + \delta_{p} \right)^{\Delta m_{ij}},$$

(10.7)

where the fractionation factor is

$$\delta_{p} = \left(\frac{R_{uv}^{N}}{R_{uv}^{M}} \right)^{1/\Delta m_{uv}} - 1.$$

(10.8)

For Sr isotopes,

$$\left(\frac{^{87}Sr}{^{86}Sr}\right)_{true} = \left(\frac{^{87}Sr}{^{86}Sr}\right)_{measured}\left(1+\delta_p\right)^{(87-86)}$$

$$= \left(\frac{^{87}Sr}{^{86}Sr}\right)_{measured}\left(1+\delta_p\right), \qquad (10.9)$$

where

$$\delta_p = \left[\left(\frac{^{86}Sr}{^{88}Sr}\right)_{true}\bigg/\left(\frac{^{86}Sr}{^{88}Sr}\right)_{measured}\right]^{1/(86-88)} -1$$

$$= \left[0.1194\bigg/\left(\frac{^{86}Sr}{^{88}Sr}\right)_{measured}\right]^{-1/2} -1. \qquad (10.10)$$

For Nd isotopes,

$$\left(\frac{^{143}Nd}{^{144}Nd}\right)_{true} = \left(\frac{^{143}Nd}{^{144}Nd}\right)_{measured}\left(1+\delta_p\right)^{(143-144)}$$

$$= \left(\frac{^{143}Nd}{^{144}Nd}\right)_{measured}\left(1+\delta_p\right)^{-1}, \qquad (10.11)$$

where

$$\delta_p = \left[\left(\frac{^{146}Nd}{^{144}Nd}\right)_{true}\bigg/\left(\frac{^{146}Nd}{^{144}Nd}\right)_{measured}\right]^{1/(146-144)} -1$$

$$= \left[0.7219\bigg/\left(\frac{^{146}Nd}{^{144}Nd}\right)_{measured}\right]^{1/2} -1. \qquad (10.12)$$

The power law can be rewritten as

$$\frac{\ln\left(R_{ij}^C/R_{ij}^M\right)}{\ln\left(R_{uv}^N/R_{uv}^M\right)} = \frac{\Delta m_{ij}}{\Delta m_{uv}}. \qquad (10.13)$$

In the $\ln\left(R_{ij}^C/R_{ij}^M\right)$ vs. $\ln\left(R_{uv}^N/R_{uv}^M\right)$ plot, the data display a linear relationship with slope being $\Delta m_{ij}/\Delta m_{uv}$. Power law can also be expressed as

$$\left(R_{ij}^C/R_{ij}^M\right) = \left(R_{uv}^N/R_{uv}^M\right)^{\frac{\Delta m_{ij}}{\Delta m_{uv}}}. \tag{10.14}$$

10.1.3. Exponential law

For exponential law, the corrected ratio is given by

$$R_{ij}^C = R_{ij}^M \left(\frac{m_i}{m_j}\right)^{\beta}, \tag{10.15}$$

where

$$\beta = \ln\left(\frac{R_{uv}^N}{R_{uv}^M}\right) \Big/ \ln\left(\frac{m_u}{m_v}\right). \tag{10.16}$$

For Sr isotopes,

$$\left(\frac{^{87}Sr}{^{86}Sr}\right)_{true} = \left(\frac{^{87}Sr}{^{86}Sr}\right)_{measured} \left(\frac{87}{86}\right)^{\beta}, \tag{10.17}$$

where

$$\begin{aligned}
\beta &= \ln\left[\left(\frac{^{86}Sr}{^{88}Sr}\right)_{true} \Big/ \left(\frac{^{86}Sr}{^{88}Sr}\right)_{measured}\right] \Big/ \ln\left(\frac{86}{88}\right) \\
&= \ln\left[0.1194 \Big/ \left(\frac{^{86}Sr}{^{88}Sr}\right)_{measured}\right] \Big/ \ln\left(\frac{86}{88}\right).
\end{aligned} \tag{10.18}$$

For Nd isotopes,

$$\left(\frac{^{143}Nd}{^{144}Nd}\right)_{true} = \left(\frac{^{143}Nd}{^{144}Nd}\right)_{measured} \left(\frac{143}{144}\right)^{\beta}, \tag{10.19}$$

where

$$\beta = \ln\left[\left(\frac{^{146}Nd}{^{144}Nd}\right)_{true}\bigg/\left(\frac{^{146}Nd}{^{144}Nd}\right)_{measured}\right]\bigg/\ln\left(\frac{146}{144}\right)$$

$$= \ln\left[0.7219\bigg/\left(\frac{^{146}Nd}{^{144}Nd}\right)_{measured}\right]\bigg/\ln\left(\frac{146}{144}\right). \tag{10.20}$$

The exponential law can also be presented as

$$R_{ij}^C\big/R_{ij}^M = \left(R_{uv}^N\big/R_{uv}^M\right)^{\frac{\ln(m_i/m_j)}{\ln(m_u/m_v)}}, \tag{10.21}$$

which can be re-expressed as

$$\frac{\ln\left(R_{ij}^C\big/R_{ij}^M\right)}{\ln\left(R_{uv}^N\big/R_{uv}^M\right)} = \frac{\ln\left(m_i/m_j\right)}{\ln\left(m_u/m_v\right)}. \tag{10.22}$$

In the $\ln\left(R_{ij}^C\big/R_{ij}^M\right)$ vs. $\ln\left(R_{uv}^N\big/R_{uv}^M\right)$ plot, the data display a linear relationship with slope being $\dfrac{\ln\left(m_i/m_j\right)}{\ln\left(m_u/m_v\right)}$ for exponential law. In comparison, according to Eq. (10.13), the slope in the $\ln\left(R_{ij}^C\big/R_{ij}^M\right)$ vs. $\ln\left(R_{uv}^N\big/R_{uv}^M\right)$ plot is $\Delta m_{ij}\big/\Delta m_{uv}$ for power law. The difference in slopes between power law and exponential law is very small for heavy isotopic ratios. For example, for the Sr isotopic ratios, power law gives the slope as

$$\frac{\ln\left[\left(\frac{^{87}Sr}{^{86}Sr}\right)_{true}\bigg/\left(\frac{^{87}Sr}{^{86}Sr}\right)_{measured}\right]}{\ln\left[\left(\frac{^{86}Sr}{^{88}Sr}\right)_{true}\bigg/\left(\frac{^{86}Sr}{^{88}Sr}\right)_{measured}\right]} = \frac{87-86}{86-88} = -0.5, \tag{10.23}$$

and the exponential law yields the slope as

$$\frac{\ln\left[\left(\dfrac{^{87}Sr}{^{86}Sr}\right)_{true} \Big/ \left(\dfrac{^{87}Sr}{^{86}Sr}\right)_{measured}\right]}{\ln\left[\left(\dfrac{^{86}Sr}{^{88}Sr}\right)_{true} \Big/ \left(\dfrac{^{86}Sr}{^{88}Sr}\right)_{measured}\right]} = \frac{\ln\left(\dfrac{87}{86}\right)}{\ln\left(\dfrac{86}{88}\right)} = -0.50287.$$

$$(10.24)$$

Example. Calculate the normalized $^{87}Sr/^{86}Sr$ when then measured $^{87}Sr/^{86}Sr = 0.704333$ and measured $^{86}Sr/^{88}Sr = 0.11951$ using linear, power and exponential laws, respectively. For Sr, the universal choice is $^{86}Sr/^{88}Sr = 0.1194$.

The mass fractionation factor from linear law is

$$\delta = \frac{1}{2}\left[1 - \frac{0.1194}{\left(^{86}Sr/^{88}Sr\right)_{measured}}\right] = \frac{1}{2}\left[1 - \frac{0.1194}{0.11951}\right] = 0.00046,$$

thus

$$\left(\frac{^{87}Sr}{^{86}Sr}\right)_{true} = \left(\frac{^{87}Sr}{^{86}Sr}\right)_{measured}(1+\delta)$$

$$= 0.704333(1 + 0.00046) = 0.704657.$$

The mass fractionation factor from power law is

$$\delta_p = \left[0.1194 \Big/ \left(\frac{^{86}Sr}{^{88}Sr}\right)_{measured}\right]^{-1/2} - 1$$

$$= [0.1194/0.11951]^{-1/2} - 1 = 0.000461,$$

thus

$$\left(\frac{^{87}Sr}{^{86}Sr}\right)_{true} = \left(\frac{^{87}Sr}{^{86}Sr}\right)_{measured}(1+\delta_p)$$

$$= 0.704333(1 + 0.000461) = 0.704657.$$

For exponential law, we have the fractionation factor

$$\beta = \ln \left[0.1194 \middle/ \left(\frac{^{86}Sr}{^{88}Sr} \right)_{measured} \right] \middle/ \ln \left(\frac{86}{88} \right)$$

$$= \ln \left(\frac{0.1194}{0.11951} \right) \middle/ \ln \left(\frac{86}{88} \right)$$

$$= 0.040055,$$

thus

$$\left(\frac{^{87}Sr}{^{86}Sr} \right)_{true} = \left(\frac{^{87}Sr}{^{86}Sr} \right)_{measured} \left(\frac{87}{86} \right)^{\beta}$$

$$= 0.704333 \left(\frac{87}{86} \right)^{0.040055} = 0.704659.$$

Note that the normalized (true) values on the basis of three laws are very similar to each other in the example.

Example. Calculate the normalized $^{143}Nd/^{144}Nd$ when then measured $^{143}Nd/^{144}Nd = 0.512776$ and measured $^{146}Nd/^{144}Nd = 0.72175$ using linear, power and exponential laws, respectively. For Nd isotope, the choice for normalization is $^{146}Nd/^{144}Nd = 0.7219$.

For linear law, we have the fractionation factor:

$$\delta = \frac{1}{2} \left[\frac{0.7219}{\left(^{146}Nd/^{144}Nd \right)_{measured}} - 1 \right]$$

$$= \frac{1}{2} \left(\frac{0.7219}{0.72175} - 1 \right) = 0.000104,$$

thus

$$\left(\frac{^{143}Nd}{^{144}Nd} \right)_{true} = \left(\frac{^{143}Nd}{^{144}Nd} \right)_{measured} (1 - \delta)$$

$$= 0.512776(1 - 0.000104) = 0.512723.$$

The fractionation factor for power law is

$$\delta_p = \left[0.7219 \middle/ \left(\frac{^{146}Nd}{^{144}Nd} \right)_{measured} \right]^{1/2} - 1$$

$$= \left(\frac{0.7219}{0.72175} \right)^{1/2} - 1 = 0.000104,$$

thus the true value is

$$\left(\frac{^{143}Nd}{^{144}Nd} \right)_{true} = \left(\frac{^{143}Nd}{^{144}Nd} \right)_{measured} \left(1 + \delta_p \right)^{-1}$$

$$= 0.512776 \left(1 + 0.000104 \right)^{-1} = 0.512723.$$

For exponential law, we have

$$\beta = \ln \left[0.7219 \middle/ \left(\frac{^{146}Nd}{^{144}Nd} \right)_{measured} \right] \middle/ \ln \left(\frac{146}{144} \right)$$

$$= \ln \left[0.7219/0.72175 \right] \middle/ \ln \left(\frac{146}{144} \right) = 0.0150657,$$

thus the true value is

$$\left(\frac{^{143}Nd}{^{144}Nd} \right)_{true} = \left(\frac{^{143}Nd}{^{144}Nd} \right)_{measured} \left(\frac{143}{144} \right)^{\beta}$$

$$= 0.512776 \left(\frac{143}{144} \right)^{0.0150657} = 0.512722.$$

10.2. Internal and External Correction of Mass Fractionation

10.2.1. Internal correction

Although internal correction does not need any external isotopic standards, the correction requires that the element has at least two natural non-radiogenic isotopes. Some elements, e.g., Nd and Sr, satisfy this requirement. Take Sr and Nd internal linear correction as examples.

From a linear law or a linear approximation of the exponential law, the magnitude of the mass bias can be inferred from the measurement of this ratio in the sample

$$(^{86}Sr/\,^{88}Sr)^{spl}_{true} = 0.1194 = (^{86}Sr/\,^{88}Sr)^{spl}_{measured}(1-2\delta^{Sr}_{spl}) \ . \ (10.25)$$

And δ^{spl}_{Sr} calculated from Eq. (10.25) is used to normalize the $^{87}Sr/^{86}Sr$ to correct for mass bias

$$(^{87}Sr/\,^{86}Sr)^{spl}_{normalized} = (^{87}Sr/\,^{86}Sr)^{spl}_{measured}\left(1+\delta^{spl}_{Sr}\right) \ . \quad (10.26)$$

For Nd, the common choice is $^{146}Nd/^{144}Nd=0.7219$. From a linear approximation, the magnitude of the mass bias can be inferred from the measurement of this ratio

$$(^{146}Nd/\,^{144}Nd)^{spl}_{true} = 0.7219 = (^{146}Nd/\,^{144}Nd)^{spl}_{measured}(1+2\delta^{spl}_{Nd}). \quad (10.27)$$

And δ^{spl}_{Nd} calculated from Eq. (10.27) is used to normalize the $^{143}Nd/^{144}Nd$ to correct for mass bias

$$(^{143}Nd/\,^{144}Nd)^{spl}_{normalized} = (^{143}Nd/\,^{144}Nd)^{spl}_{measured}(1-\delta^{spl}_{Nd}) \ . \ (10.28)$$

10.2.2. External correction

By using internal correction, isotopic ratios can be obtained at a very high precision (for example, 10 ppm for $^{87}Sr/^{86}Sr$ and $^{143}Nd/^{144}Nd$ by mass spectrometers). Some elements, however, do not have two natural non-radiogenic isotopes. For example, Pb has three radiogenic isotopes (^{206}Pb, ^{207}Pb and ^{208}Pb) but only one stable isotope ^{204}Pb. We cannot use the internal correction of mass fractionation. But we can use an external Pb isotopic standard to correct the mass fractionation, assuming that the mass fractionation factor for the external standard (δ^{std}_{Pb}) is identical to that for the sample δ^{spl}_{Pb}. For the Pb isotopic standard NBS981, we have

$$(^{206}Pb/\,^{204}Pb)^{std}_{true} = 16.9356 = (^{206}Pb/\,^{204}Pb)^{std}_{meas}(1+2\delta^{std}_{Pb}). \quad (10.29)$$

Note that 16.9356 is the true $^{206}Pb/^{204}Pb$ for NBS 981 (Todt et al., 1996). For a natural sample, we have

$$({}^{206}\text{Pb} / {}^{204}Pb)_{true}^{spl} = ({}^{206}\text{Pb} / {}^{204}Pb)_{meas}^{spl}(1 + 2\delta_{Pb}^{spl}). \qquad (10.30)$$

From (10.29) we get

$$\delta_{Pb}^{std} = \frac{1}{2}\left(\frac{16.9356}{({}^{206}\text{Pb} / {}^{204}Pb)_{meas}^{std}} - 1\right). \qquad (10.31)$$

Assuming $\delta_{Pb}^{spl} = \delta_{Pb}^{std}$, substitution of (10.31) into (10.30) yields the true ${}^{206}\text{Pb}/{}^{204}\text{Pb}$ ratio in the sample.

Example. Suppose we obtain the following results for standard NBS 981 and a sample: $({}^{206}\text{Pb} / {}^{204}Pb)_{meas}^{std} = 16.891$, $({}^{206}\text{Pb} / {}^{204}Pb)_{meas}^{std} = 18.712$. Find the true ratio, $({}^{206}\text{Pb} / {}^{204}Pb)_{true}^{spl}$, of the sample.
From (10.31) we have

$$\delta_{Pb}^{std} = \frac{1}{2}\left(\frac{16.9356}{16.891} - 1\right) = 0.00132.$$

Assuming $\delta_{Pb}^{spl} = \delta_{Pb}^{std} = 0.00132$ and from (10.30), we obtain the true ratio in the sample:

$$({}^{206}\text{Pb} / {}^{204}Pb)_{true}^{spl} = 18.712(1 + 2 \times 0.00132) = 18.761.$$

Another way to do the correction of Pb isotope mass fractionation is to use double spike method, which will be discussed in Chapter 11.

10.3. Dynamic Measurements of Nd and Sr Isotopic Compositions

Multi-collection mass spectrometers can analyze isotope ratios in a static mode to eliminate the errors from beam instability. However, the static multi-collection method depends on the extent to which the collectors (e.g., Faraday cups) are identical and to the extent to which the gain of each collector is stable. An alternative approach is to use the so-called "dynamic" multi-collector mode, to cancel out beam instability, detector bias, and performing a power-law mass fractionation correction. The following descriptions are modified from the Finnigan MAT 262 Operating manual (Finnigan, 1992).

10.3.1. Sr double collector analysis

Faraday cup number   6   5   4   3

1st Measurement: Mass   87   88

2nd Measurement: Mass   (85)   86   87

From the power law Eq. (10.7), we have the measured isotopic ratio as

$$R_{ij}^{M} = R_{ij}^{C}\left(1+\delta_{p}\right)^{-\Delta m_{ij}}.$$

The first scan uses Faraday cup 3 to measure ^{88}Sr and cup 4 to measure ^{87}Sr. On the basis of the cup efficiency factor (K_3 for cup 3 and K_4 for cup 4), the measured ^{88}Sr/^{87}Sr ratio is

$$R_1 = \left(\frac{^{88}\text{Sr}}{^{87}\text{Sr}}\right)_{measured} = \left(\frac{^{88}\text{Sr}}{^{87}\text{Sr}}\right)_{true}\frac{K_3}{K_4}\left(1+\delta\right)^{-\Delta m_{ij}}$$

$$= \left(\frac{^{88}\text{Sr}}{^{87}\text{Sr}}\right)_{true}\frac{K_3}{K_4}\left(1+\delta\right)^{-(88-87)} \qquad (10.32)$$

$$= \left(\frac{^{88}\text{Sr}}{^{87}\text{Sr}}\right)_{true}\frac{K_3}{K_4}\left(1+\delta\right)^{-1}.$$

Similarly, with the second scan, the measured ^{87}Sr/^{86}Sr ratio is related to its true ratio by

$$R_2 = \left(\frac{^{87}\text{Sr}}{^{86}\text{Sr}}\right)_{measured} = \left(\frac{^{87}\text{Sr}}{^{86}\text{Sr}}\right)_{true}\frac{K_3}{K_4}\left(1+\delta\right)^{-1}. \qquad (10.33)$$

Dividing (10.33) by (10.32), we obtain

$$\frac{R_2}{R_1} = \frac{\left(^{87}\text{Sr}/^{86}\text{Sr}\right)_{measured}}{\left(^{88}\text{Sr}/^{87}\text{Sr}\right)_{measured}} = \left(\frac{^{87}\text{Sr}}{^{86}\text{Sr}}\right)_{true}^{2}\left(\frac{^{86}\text{Sr}}{^{88}\text{Sr}}\right)_{true}. \qquad (10.34)$$

Note that mass fractionation factor δ for power law and the cup efficiency factor K are completely cancelled out in (10.34). However, readers may find that if we use linear law or exponential law, the fractionation factors cannot be completely eliminated by division.

Since the true ^{86}Sr/^{88}Sr is 0.1194, we have

$$\frac{R_2}{R_1} = 0.1194 \left(\frac{^{87}\text{Sr}}{^{86}\text{Sr}} \right)^2_{true}.$$

Thus the true (normalized) ^{87}Sr/^{86}Sr is

$$\left(\frac{^{87}\text{Sr}}{^{86}\text{Sr}} \right)_{true} = \sqrt{\frac{R_2}{0.1194 R_1}}. \qquad (10.35)$$

10.3.2. Nd double collector analysis

Faraday cup number		6	5
1st Measurement:	Mass	143	144
2nd Measurement:	Mass	144	145
3rd Measurement:	Mass	145	146

For the first scan, the measured ^{143}Nd/^{144}Nd is given by

$$
\begin{aligned}
R_1 &= \left(\frac{^{143}\text{Nd}}{^{144}\text{Nd}} \right)_{measured} = \left(\frac{^{143}\text{Nd}}{^{144}\text{Nd}} \right)_{true} \frac{K_6}{K_5} (1+\delta)^{-\Delta m_{ij}} \\
&= \left(\frac{^{143}\text{Nd}}{^{144}\text{Nd}} \right)_{true} \frac{K_6}{K_5} (1+\delta)^{-(143-144)} \qquad (10.36) \\
&= \left(\frac{^{143}\text{Nd}}{^{144}\text{Nd}} \right)_{true} \frac{K_6}{K_5} (1+\delta),
\end{aligned}
$$

where K_5 and K_6 are the Faraday cup efficiency factors for cup 5 and 6, respectively, and δ is the mass fractionation factor per mass unit. Similarly, for the second measurement (scan), we have

$$R_2 = \left(\frac{^{145}\text{Nd}}{^{144}\text{Nd}} \right)_{measured} = \left(\frac{^{145}\text{Nd}}{^{144}\text{Nd}} \right)_{true} \frac{K_5}{K_6} (1+\delta)^{-1}, \qquad (10.37)$$

and for the third measurement, we have

$$R_3 = \left(\frac{^{146}\text{Nd}}{^{145}\text{Nd}}\right)_{measured} = \left(\frac{^{146}\text{Nd}}{^{145}\text{Nd}}\right)_{true} \frac{K_5}{K_6}(1+\delta)^{-1}. \quad (10.38)$$

Combination of (10.37) and (10.38) yields

$$\sqrt{R_2 R_3} = \sqrt{\left(\frac{^{146}\text{Nd}}{^{144}\text{Nd}}\right)_{true}} \frac{K_5}{K_6}(1+\delta)^{-1}. \quad (10.39)$$

Combination of (10.36) and (10.39) gives

$$R_1\sqrt{R_2 R_3}$$

$$= \left[\left(\frac{^{143}\text{Nd}}{^{144}\text{Nd}}\right)_{true} \frac{K_6}{K_5}(1+\delta)\right]\left(\sqrt{\left(\frac{^{146}\text{Nd}}{^{144}\text{Nd}}\right)_{true}} \frac{K_5}{K_6}(1+\delta)^{-1}\right)$$

$$= \left(\frac{^{143}\text{Nd}}{^{144}\text{Nd}}\right)_{true} \sqrt{\left(\frac{^{146}\text{Nd}}{^{144}\text{Nd}}\right)_{true}},$$

or

$$\left(\frac{^{143}\text{Nd}}{^{144}\text{Nd}}\right)_{true} = R_1\sqrt{R_2 R_3} \Bigg/ \sqrt{\left(\frac{^{146}\text{Nd}}{^{144}\text{Nd}}\right)_{true}}.$$

In this way, the cup efficiency factor K and the fractionation factor per mass unit (δ) are eliminated.

Since the true $^{146}\text{Nd}/^{144}\text{Nd} = 0.7219$, the true $^{143}\text{Nd}/^{144}\text{Nd}$ value is

$$\left(\frac{^{143}\text{Nd}}{^{144}\text{Nd}}\right)_{true} = R_1\sqrt{R_2 R_3 / 0.7219}. \quad (10.40)$$

10.3.3. Nd triple collector analysis

Faraday cup number		7	6	5
1st Measurement:	Mass	143	144	145
2nd Measurement:	Mass	144	145	146

Initially, ratios R_1 and R_2 are measured in the first scan, where

$$R_1 = \left(\frac{^{143}\text{Nd}}{^{144}\text{Nd}}\right)_{measured} = \left(\frac{^{143}\text{Nd}}{^{144}\text{Nd}}\right)_{true} \frac{K_7}{K_6}(1+\delta), \qquad (10.41)$$

$$R_2 = \left(\frac{^{145}\text{Nd}}{^{144}\text{Nd}}\right)_{measured} = \left(\frac{^{145}\text{Nd}}{^{144}\text{Nd}}\right)_{true} \frac{K_5}{K_6}(1+\delta)^{-1}. \qquad (10.42)$$

K_5, K_6 and K_7 are cup efficiency factors for cup 5, 6 and 7, respectively, δ is the fractionation factor per mass unit.
In the second measurement, we have:

$$R_3 = \left(\frac{^{145}\text{Nd}}{^{144}\text{Nd}}\right)_{measured} = \left(\frac{^{145}\text{Nd}}{^{144}\text{Nd}}\right)_{true} \frac{K_6}{K_7}(1+\delta)^{-1}, \qquad (10.43)$$

$$R_4 = \left(\frac{^{146}\text{Nd}}{^{145}\text{Nd}}\right)_{measured} = \left(\frac{^{146}\text{Nd}}{^{145}\text{Nd}}\right)_{true} \frac{K_5}{K_6}(1+\delta)^{-1}. \qquad (10.44)$$

Multiplying R_1 by R_3 can eliminate K_6, K_7 and δ:

$$R_1 R_3 = \left(\frac{^{143}\text{Nd}}{^{144}\text{Nd}}\right)_{true} \left(\frac{^{145}\text{Nd}}{^{144}\text{Nd}}\right)_{true}. \qquad (10.45)$$

And dividing R_4 by R_2 can cancel out K_5, K_6 and δ:

$$\frac{R_4}{R_2} = \left(\frac{^{146}\text{Nd}}{^{145}\text{Nd}}\right)_{true} \left(\frac{^{144}\text{Nd}}{^{145}\text{Nd}}\right)_{true}. \qquad (10.46)$$

Combination of (10.45) and (10.46) leads to

$$R_1 R_3 \sqrt{R_4 / R_2} = \left(\frac{^{143}\text{Nd}}{^{144}\text{Nd}}\right)_{true} \left(\frac{^{145}\text{Nd}}{^{144}\text{Nd}}\right)_{true} \sqrt{\left(\frac{^{146}\text{Nd}}{^{145}\text{Nd}}\right)_{true} \left(\frac{^{144}\text{Nd}}{^{145}\text{Nd}}\right)_{true}}$$

$$= \left(\frac{^{143}\text{Nd}}{^{144}\text{Nd}}\right)_{true} \sqrt{\left(\frac{^{146}\text{Nd}}{^{144}\text{Nd}}\right)_{true}}.$$

Since the true value $^{146}\text{Nd}/^{144}\text{Nd} = 0.7219$, we obtain

$$\left(\frac{^{143}\text{Nd}}{^{144}\text{Nd}}\right)_{true} = R_1 R_3 \sqrt{\frac{R_4}{0.7219 R_2}} \, . \qquad (10.47)$$

References

Finnigan, 1992. Finnigan MAT 262 Operating Manual, Chapter 8, p. 1-17.

Lee, C.T., Yin, Q.Z. and Lee, T.C., 2001. An internal normalization technique for unmixing total-spiked mixtures with application to MC-ICP-MS. Computers & Geosciences, 27: 577-581.

Todt, W., Cliff, R.A., Hanser, A. and Hofmann, A.W., 1996. Evaluation of a ^{202}Pb-^{205}Pb double spike for high-precision lead isotope analysis. In: A.R. Basu, Hart, S. R. (Editor), Earth processes: reading the isotopic code. Geophysical Monograph Am. Geophys. Union, pp. 429-437.

Wasserburg, G.J., Jacobsen, S.B., DePaolo, D.J., McCulloch, M.T. and Wen, T., 1981. Precise determination of Sm/Nd ratios, Sm and Nd isotopic abundances in standard solutions. Geochim. Cosmochim. Acta, 45: 2311-2323.

Chapter 11

Isotope Dilution

The isotope dilution method consists of mixing a natural sample with an artificial spike and measuring the isotopic ratios of the mixture using mass spectrometry, providing very precise quantitative determination of the concentrations of elements in trace quantities. A spike is a solution that contains a known concentration of the element, artificially enriched in one of its minor isotopes. For example, natural Rb samples have 27.84% ^{87}Rb and 72.16% ^{85}Rb (Fig. 11.1A). Rb spikes are made by artificially enrich the minor isotope ^{87}Rb. And a solution with 90% ^{87}Rb and 10% ^{85}Rb (Fig. 11.1B) is a Rb spike. Of course, a solution with 99.99% ^{87}Rb and 0.01% ^{85}Rb is a better Rb spike. When the known quantities (mass) of the sample and spike are mixed, the resulting isotopic compositions can be used to calculate the concentration of the element in the sample.

The spike in most cases is single spike, but it can be double spike. Single spike method provides the information of the concentration of the sample, whereas the double-spike method can yield the concentration of the sample and the mass fractionation factor that can be used to calculate the true isotopic ratios in the sample. Take Pb isotope as an example. Natural Pb samples have 1.4% ^{204}Pb, 24.4% ^{206}Pb, 22.1% ^{207}Pb and 52.4% ^{208}Pb. A single spike is made by artificially concentrating one of the minor Pb isotopes, for example ^{204}Pb. A double spike is made by artificially concentrating two minor isotopes. For example, we can concentrate ^{204}Pb and ^{206}Pb isotopes to make ^{204}Pb-^{206}Pb double spike or concentrate ^{204}Pb and ^{207}Pb to make ^{204}Pb-^{207}Pb double spike.

The theory of isotope dilution is a fine example that a better understanding of the theory can improve experimental measurements.

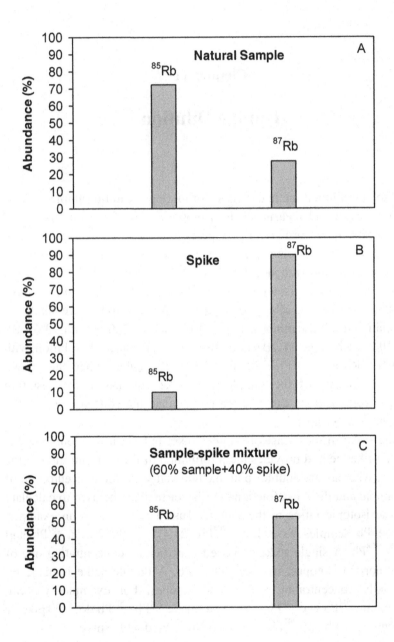

Fig. 11.1. Abundances of ^{87}Rb and ^{85}Rb in (A) a natural sample, (B) a ^{87}Rb spike, and (C) a sample-spike mixture.

11.1. Single Spike Method

11.1.1. Single spike equation using isotope abundances

Single spiking method is essentially a problem of two-component mixing. The ratio of the abundances of two isotopes A and B of an element in the mixture is (Faure, 1986)

$$R_M = \frac{Ab_N^A N + Ab_S^A S}{Ab_N^B N + Ab_S^B S},$$ (11.1)

where Ab_N^A and Ab_N^B are the isotope abundances of isotopes A and B, respectively, in the natural sample; Ab_S^A and Ab_S^B are the isotope abundances of A and B, respectively, in the spike; N and S are the total atoms of the normal element and of the spike, respectively. Equation (11.1) may be written as

$$N = S \left[\frac{Ab_S^A - R_m Ab_S^B}{R_m Ab_N^B - Ab_N^A} \right].$$ (11.2)

Note that for naturally occurring non-radiogenic isotopes, for example, ^{84}Sr and ^{88}Sr, their natural abundances Ab_N^A and Ab_N^B are known in the natural sample. The spike Ab_S^A and Ab_S^B are also known. Thus one measurement of R_M in the mixture is enough to calculate the number of atoms of the element N in the sample. Since

$$N = \frac{C_N M_N}{W_N} A ,$$ (11.3)

and

$$S = \frac{C_{Spike} M_S}{W_S} A ,$$ (11.4)

where A is the Avogadro's number. W_N and W_S are the atomic weights of the element in the sample and spike, respectively. M_N and M_S are the weight of the sample and the weight of the spike, respectively; C_N and C_{Spike} are the concentration of the element in the normal sample and spike, respectively. Thus, $C_N M_N$ is the weight of the element from the normal sample and $C_{Spike} M_S$ is the weight of the element from the spike.

Substitution of (11.3) and (11.4) into (11.2) gives the concentration of the element in the natural sample

$$C_N = C_{Spike} \frac{M_S}{M_N} \frac{W_N}{W_S} \left[\frac{Ab_S^A - R_m Ab_S^B}{R_m Ab_N^B - Ab_N^A} \right]. \qquad (11.5)$$

In the above equation, the isotopic composition in the mixture R_m is measured by amass spectrometer, the weight of the normal sample (M_N) and the weight of the spike (M_S) can be measured by a analytical weighing balance, and all other parameters (C_{Spike}, W_N, Ab_N^A, Ab_N^B, Ab_S^A, Ab_S^B) are known. Thus, C_N can be calculated from the above equation.

Example. Suppose we add 0.0681 gram of Rb spike to 0.03893 g natural rock. In the spiked mixture, mass spectrometric measurement gives $^{85}Rb/^{87}Rb=1.162$ and $^{84}Sr/^{86}Sr=0.1043$. The information about the spike and natural samples are provided in Table 11.1. Calculate the Rb concentrations in the natural sample.

Table 11.1. The ^{85}Rb and ^{87}Rb abundances in a spike and natural samples, and Rb concentration in the spike.

	Rb concentration	^{87}Rb abundance	^{85}Rb abundance	Rb atomic weight
Spike	9.831 ppm	0.97972	0.02028	86.869
Sample	? ppm	0.2784	0.7216	85.468

Now we have the Rb concentration in the spike $C_{Spike} = 9.831$, the weight of the spike $M_S = 0.0681$ g, the weight of the natural sample $M_N = 0.03893$ g, the atomic mass of the spike $W_S = 86.869$, the atomic mass of the natural sample $W_N = 85.468$, the abundance of ^{87}Rb in the spike $Ab_S^A = 0.97972$, the abundance of ^{85}Rb in the spike $Ab_S^B = 0.02028$, the abundance of ^{87}Rb in the natural sample $Ab_N^A = 0.2784$, the abundance of ^{85}Rb in the natural sample $Ab_N^B = 0.7216$, and ^{87}Rb/^{85}Rb in the mixture $R_m = 0.8606$. Substituting all the above parameters into Eq. (11.5), we obtain

$$C_N^{Rb} = 9.831 \times \frac{0.0681}{0.03893} \times \frac{85.468}{86.869} \times \left(\frac{0.97972 - 0.8606 \times 0.02028}{0.8606 \times 0.7216 - 0.2784} \right)$$

$$= 49.09 \text{ ppm.}$$

Example. Calculate the Sr concentration in the natural sample. The weight of the Sr spike is still $M_S = 0.03893 \text{ g}$, the weight of the sample is still $M_N = 0.0681 \text{ g}$, and the measured ^{84}Sr/^{88}Sr in the mixture is $R_m^{Sr} = 0.953$. The Sr spike information is provided in Table 11.2.

Table 11.2. The ^{84}Sr and ^{88}Sr abundances in a spike and natural samples, and Sr concentration in the spike.

Sr	^{84}Sr	^{88}Sr	Sr atomic
	concentration	abundance	abundance	weight
Spike	14.78 ppm	0.8246	0.1233	84.525
Sample	? ppm	0.0056	0.8258	87.62

Now we have the Sr concentration in the spike $C_{Spike} = 14.78$, the weight of the spike $M_S = 0.0681 \text{ g}$, the weight of the natural sample $M_N = 0.03893 \text{ g}$, the atomic mass of the spike $W_S = 84.525$, the atomic mass of the natural sample $W_N = 87.62$, the abundance of ^{84}Sr in the spike $Ab_S^A = 0.8246$, the abundance of ^{88}Sr in the spike $Ab_S^B = 0.1233$, the abundance of ^{84}Sr in the natural sample $Ab_N^A = 0.0056$, the abundance of ^{88}Sr in the natural sample $Ab_N^B = 0.8258$, and ^{84}Sr/^{88}Sr in the mixture $R_m = 0.953$. Substituting all the above parameters into Eq. (11.5), we obtain

$$C_N^{Sr} = 14.78 \times \frac{0.0681}{0.03893} \times \frac{87.62}{84.525} \times \left(\frac{0.8246 - 0.953 \times 0.1233}{0.953 \times 0.8258 - 0.0056} \right)$$

$$= 24.25 \text{ ppm.}$$

Example. Calculate the U concentration in the natural sample. Suppose the weight of the U spike is $M_S = 0.0166$ g, the weight of the sample is $M_N = 0.055$ g, and the measured $^{233}U/^{238}U$ in the mixture is $R_m^U = 3.512$. The U spike information is given in Table 11.3.

Table 11.3. The ^{233}U and ^{238}U abundances in a spike and natural samples, and U concentration in the spike.

	U concentration	^{233}U abundance	^{238}U abundance	U atomic weight
Spike	2.051 ppm	0.99491	0.00229	233.45
Sample	? ppm	0	0.992	238.03

Now we have the U concentration in the spike $C_{Spike} = 2.051$, the weight of the spike $M_S = 0.0166$ g, the weight of the natural sample $M_N = 0.055$ g, the atomic mass of the spike $W_S = 233.45$, the atomic mass of the natural sample $W_N = 238.03$, the abundance of ^{233}U in the spike $Ab_S^A = 0.99491$, the abundance of ^{238}U in the spike $Ab_S^B = 0.003$, the abundance of ^{233}U in the natural sample $Ab_N^A = 0$, the abundance of ^{238}U in the natural sample $Ab_N^B = 0.992$, and the measured $^{233}U/^{238}U$ in the mixture $R_m = 0.953$. By substituting all the above parameters into Eq. (11.5), we obtain

$$C_N^U = 2.051 \times \frac{0.055}{0.0166} \times \frac{238.03}{233.45} \times \left(\frac{0.99491 - 3.512 \times 0.00229}{3.512 \times 0.99491 - 0} \right)$$

$$= 1.963 \text{ ppm.}$$

Example. Calculate the Th concentration in a natural sample. Suppose the weight of the Th spike is $M_S = 0.0166$ g, the weight of the sample is $M_N = 0.055$ g, and the measured ^{229}Th/^{232}Th in the mixture is $R_m^{Th} = 0.115$. The Th spike information is provided in Table 11.4.

Table 11.4. The ^{229}Th and ^{232}Th abundances in a spike and natural samples, and Th concentration in the spike.

	Th concentration	^{229}Th abundance	^{232}Th abundance	Th atomic weight
Spike	0.314 ppm	0.997	0.003	229.35
Sample	? ppm	0	1.0	232.04

Now we have the Th concentration in the spike $C_{Spike} = 0.314$, the weight of the spike $M_S = 0.0166$ g, the weight of the natural sample $M_N = 0.055$ g, the atomic mass of the spike $W_S = 229.35$, the atomic mass of the natural sample $W_N = 232.04$, the abundance of ^{229}Th in the spike $Ab_S^A = 0.997$, the abundance of ^{232}Th in the spike $Ab_S^B = 0.003$, the abundance of ^{229}Th in the natural sample $Ab_N^A = 0$, the abundance of ^{232}Th in the natural sample $Ab_N^B = 1.0$, and the measured ^{229}Th/^{232}Th in the mixture $R_m = 0.115$. By substituting all the above parameters into Eq. (11.5), we obtain

$$C_N^{Th} = 0.314 \times \frac{0.055}{0.0166} \times \frac{232.04}{229.35} \times \left(\frac{0.997 - 0.115 \times 0.003}{0.115 \times 1.0 - 0} \right)$$
$$= 9.122 \text{ ppm.}$$

11.1.2. Single spike equation using isotope ratios

In many cases, it is more convenient to describe the isotope dilution method using isotopic ratios rather than abundances alone. R_M in Eq. (11.1) can be expressed as a function of isotopic ratios in the natural sample R_N and the spike R_S.

$$R_N = \frac{Ab_N^A}{Ab_N^B}, \qquad (11.6)$$

$$R_S = \frac{Ab_S^A}{Ab_S^B}.$$ (11.7)

Substitution of Eqs. (11.6) and (11.7) into Eq. (11.1) results in

$$R_M = \frac{Ab_N^A N + Ab_S^A S}{Ab_N^B N + Ab_S^B S}$$

$$= \frac{Ab_N^B R_N N + Ab_S^B R_S S}{Ab_N^B N + Ab_S^B S} = \frac{R_N + qR_S}{1 + q},$$ (11.8)

where

$$q = \frac{Ab_S^B S}{Ab_N^B N}.$$ (11.9)

According to Eq. (11.8), q is related to the isotopic ratios of the spike, the natural sample and the mixture by

$$q = \frac{R_M - R_N}{R_S - R_M}.$$ (11.10)

Equation (11.8) may be written as

$$R_M(1 + q) = R_N + qR_S.$$ (11.11)

Equation (11.11) is very useful for the theory of double spiking that we are going to study later on. On the basis of Eq. (11.9), we have

$$N = \frac{S}{q} \frac{Ab_S^B}{Ab_N^B}.$$ (11.12)

Substituting Eqs. (11.3) and (11.4) into Eq. (11.12), we find the concentration of the element in the normal sample

$$C_N = C_{Spike} \frac{R_S - R_M}{R_M - R_N} \frac{M_S}{M_N} \frac{W_N}{W_S} \frac{Ab_S^B}{Ab_N^B}.$$ (11.13)

Equation (11.13) is written in a manner so that the numerator and the corresponding denominator have the same dimension.

Problem. Use Eq. (11.13) to calculate the sample Rb, Sr, U and Th concentrations in the last four examples.

For Rb, since

$$R_S^{Rb} = 0.97972/0.02028 = 48.31,$$

$$R_N^{Rb} = 0.2784/0.7216 = 0.3858,$$

$$R_m^{Rb} = 0.8606,$$

we obtain from Eq. (11.13)

$$C_N^{Rb} = 9.831 \times \frac{48.31 - 0.8606}{0.8606 - 0.3858} \times \frac{0.0681}{0.03893} \times \frac{85.46787}{86.869} \times \frac{0.02028}{0.7216}$$
$$= 49.09 \text{ ppm.}$$

For Sr, since

$$R_S^{Sr} = 0.8246/0.1233 = 6.685,$$

$$R_N^{Sr} = 0.0056/0.8258 = 0.00678,$$

$$R_m^{Sr} = 0.953,$$

we have

$$C_N^{Sr} = 14.78 \times \frac{6.685 - 0.953}{0.953 - 0.00678} \times \frac{0.0681}{0.03893} \times \frac{87.62}{84.525} \times \frac{1.1233}{0.8246}$$
$$= 24.25 \text{ ppm.}$$

For U, since

$$R_S^U = 0.9949/0.00229 = 435.2,$$

$$R_N^U = 0,$$

$$R_m^U = 3.512,$$

we obtain

$$C_N^U = 2.051 \times \frac{435.2 - 3.512}{3.512 - 0} \times \frac{0.055}{0.0166} \times \frac{238.03}{233.45} \times \frac{0.00229}{0.992}$$
$$= 1.963 \text{ ppm.}$$

For Th, since

$$R_S^{Th} = 0.997/0.003 = 332.3,$$

$$R_N^{Th} = 0,$$

$$R_m^{Th} = 0.115,$$

we have

$$C_N^{Th} = 0.314 \times \frac{332.3 - 0.115}{0.115 - 0} \times \frac{0.055}{0.0166} \times \frac{232.04}{229.35} \times \frac{0.003}{0.997}$$
$$= 9.122 \text{ ppm.}$$

11.1.3. Optimization

During isotope dilution, after the spike composition is selected, then Ab_S^B is fixed. The main control for us is to the change the sample/spike ratio, which is related to the measured R_M. By differentiating C with respect to R_M in Eq. (11.13), we find

$$\frac{\partial C}{\partial R_M} = \frac{R_N - R_S}{(R_M - R_N)^2} Z, \tag{11.14}$$

where

$$Z = C_{spike} \frac{M_S}{M_N} \frac{W_N}{W_S} \frac{Ab_S^B}{Ab_N^B}. \tag{11.15}$$

Combination of Eq. (11.14) with Eq. (11.13) gives

$$\frac{\partial C}{\partial R_M} \bigg/ C = \frac{(R_N - R_S)}{(R_M - R_N)(R_S - R_M)}. \tag{11.16}$$

Rewriting Eq. (11.16), we have

$$\frac{\partial C}{C} = \frac{R_M (R_N - R_S)}{(R_M - R_N)(R_S - R_M)} \frac{\partial R_M}{R_M}, \tag{11.17}$$

or

$$\frac{\sigma_C^2}{C^2} = \left[\frac{R_M (R_N - R_S)}{(R_M - R_N)(R_S - R_M)} \right]^2 \frac{\sigma_{R_M}^2}{R_M^2}. \tag{11.18}$$

Let

$$Q = \left| \frac{R_M (R_N - R_S)}{(R_M - R_N)(R_S - R_M)} \right|, \tag{11.19}$$

so that

$$\frac{\sigma_C^2}{C^2} = Q^2 \frac{\sigma_{R_M}^2}{R_M^2}. \tag{11.20}$$

If we define R_M as the ratio of the naturally minor (or absent) isotope over the naturally major isotope (for example, $^{87}Rb/^{85}Rb$, $^{84}Sr/^{88}Sr$, $^{233}U/^{238}U$, $^{229}Th/^{232}Th$), then we have $R_S > R_M > R_N$, and the absolute operation for Q can be removed so that

$$Q = \frac{R_M (R_S - R_N)}{(R_M - R_N)(R_S - R_M)}. \tag{11.21}$$

Q is always positive and is essentially the error magnification factor from σ_{R_M}/R_M to σ_C/C.

Example. Calculate the magnification factor for the above four examples related to Rb, Sr, U and Th.
From Eq. (11.21) we obtain

$$Q_{Rb} = \frac{0.8606(48.31 - 0.3858)}{(0.8606 - 0.3858)(48.31 - 0.8606)} = 1.831,$$

$$Q_{Sr} = \frac{0.953(6.685 - 0.00678)}{(0.953 - 0.00678)(6.685 - 0.953)} = 1.173,$$

$$Q_U = \frac{3.512(435.2 - 0)}{(3.512 - 0)(435.2 - 3.512)} = 1.0008,$$

$$Q_{Th} = \frac{0.115(332.3 - 0)}{(0.115 - 0)(332.3 - 0.115)} = 1.0003.$$

Note that for U and Th measurements, the use of man-made isotopes that are no longer existing in natural samples (^{229}Th and ^{233}U) can minimize the magnification factor. In the above cases, only 0.08% and 0.03% increase in the relative uncertainties for U and Th concentrations, respectively. We can analyze the magnification factor in a more systematic way. Of course, if $R_M \cong R_N$ (the sample is highly under-spiked) or if $R_M \cong R_S$ (the sample is highly over-spiked), then the magnification parameter $Q \to \infty$ and the error propagation is huge, as can be seen from Fig. 11.2. One may anticipate that there is an optimal R_M to minimize Q.

Problem. Find the optimal R_M. The criteria to find the optimal R_M for minimum Q is to set the first derivative to be zero:

$$\partial Q / \partial R_M = 0.$$

Differentiation of Eq. (11.21) with respect to R_M yields

$$\frac{\partial Q}{\partial R_M} = \frac{(R_S - R_N)\left(R_M{}^2 - R_N R_S\right)}{(R_M - R_N)^2 (R_S - R_M)^2} = 0. \qquad (11.22)$$

Since $R_N \neq R_S \neq R_M$, in order for Eq. (11.22) to be valid, we have

$$R_M{}^2 - R_N R_S = 0, \qquad (11.23)$$

or

$$R_M = \sqrt{R_N R_S}. \qquad (11.24)$$

This neat condition provides a good guidance for the single spike method.

Example. Suppose we have $R_N = 10$ for the natural sample and $R_S = 1000$, find the optimal R_M and the magnification factor at optimal R_M.

According to Eq. (11.24), we obtain the optimal

$$R_M = \sqrt{R_N R_S} = \sqrt{10 \times 1000} = 100.$$

When $R_M = 100$, the magnification factor is

$$Q = \frac{R_M (R_S - R_N)}{(R_M - R_N)(R_S - R_M)} = \frac{100(1000 - 10)}{(100 - 10)(1000 - 100)} = 1.22.$$

If the fractional error (σ_{R_M}/R_M) for R_M is 1‰, then the fractional error (σ_C/C) for C is 1.22‰.

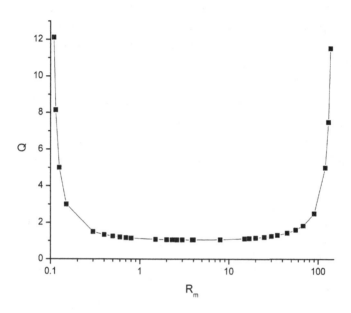

Fig. 11.2. Plot of magnification factor as a function of R_M when $R_N = 0.1$ and $R_S = 150$. The optimal value is $R_M = \sqrt{0.1 \times 150} = 3.873$. For a wide range $0.5 < R_M < 30$, the magnification factor is <1.26.

Note that the optimal condition (11.24) does not work for artificial isotope that does not exist in nature (e.g., ^{229}Th). Natural samples have $R_N = 0$, resulting in $R_m = 0$ from (11.24).
If $R_N = 0$, then Eq. (11.21) becomes

$$Q = \frac{R_S}{R_S - R_M} \,. \tag{11.25}$$

And we obtain

$$\partial Q / \partial R_M = \frac{R_S}{(R_S - R_M)^2} > 0 \tag{11.26}$$

Note that the first derivative is always positive (cannot be zero), suggesting that error magnification increases with increasing R_m when $R_N = 0$. In this case, we can add a small amount of artificial spike to the mixture (to reduce magnification factor and to save often expensive artificial spikes). As long as R_M is significantly less than R_S, the error magnification factor remains small, as can been seen from the above examples for U and Th isotope dilution).

Example. Estimate the optimal amount of Rb spike to add into the sample in the above Rb isotope dilution example from Table 11.1. The amount of natural rock is still 0.03893 g. We also have a rough estimate of the sample concentration 45 ppm from a not-so-precise measurement method.

Now we have the Rb concentration in the spike $C_{Spike} = 9.831$, the weight of the natural sample $M_N = 0.03893$ g, the atomic mass of the spike $W_S = 86.869$, the atomic mass of the natural sample $W_N = 85.468$, the abundance of ^{87}Rb in the spike $Ab_S^A = 0.97972$, the abundance of ^{85}Rb in the spike $Ab_S^B = 0.02028$, the abundance of ^{87}Rb in the natural sample $Ab_N^A = 0.2784$, the abundance of ^{85}Rb in the natural sample $Ab_N^B = 0.7216$, and ^{87}Rb/^{85}Rb in the mixture $R_m = 0.8606$. The respective isotopic ratios in the spike and the sample are

$$R_S^{Rb} = 0.97972/0.02028 = 48.31,$$

$$R_N^{Rb} = 0.2784/0.7216 = 0.3858.$$

From Eq. (11.24), the optimal R_M is

$$R_M = \sqrt{48.31 \times 0.3858} = 4.317.$$

Rewriting Eq. (11.13) for the mass of the spike, we have the optimal mass of the spike as

$$M_S = M_N \frac{C_N}{C_{Spike}} \frac{R_M - R_N}{R_S - R_M} \frac{W_S}{W_N} \frac{Ab_N^B}{Ab_S^B}. \qquad (11.27)$$

Substituting the optimal R_M and all other related parameters into (11.27), we obtain the estimated optimal mass of the spike as

$$M_S = 0.03893 \times \frac{45}{9.831} \times \frac{4.317 - 0.3858}{48.31 - 4.317} \times \frac{86.869}{85.468} \times \frac{0.7216}{0.02028}$$
$$= 0.576 \text{ g.}$$

11.2. Double Spike Method

A single spike is enriched in one minor isotope whereas a double spike is enriched in two minor isotopes (Fig. 11.3). Double spike method needs two runs. The first run is to measure the un-spiked sample and the second run is to measure the spiked mixture. Double spike method not only provides the concentrations of the sample but also corrects the mass fractionation in mass spectrometers (and thus giving the true isotopic ratios).

Dodson (1963) was the first to systematically investigate the theory of Pb isotope double spiking. The calculations related to Pb isotope double spiking may be made iteratively (Compston and Oversby, 1969) or analytically (Gale, 1970). Further development and application of the double spiking theory include the studies by Hamelin et al. (1985), Galer (1999), Johnson and Beard (1999), and Thirlwall (2000).

The fractionation curve for the unspiked sample (unknown) from the linear mass fractionation law is described by

$$x_U = X_U^* \left(1 + \alpha_X \delta_U\right) \tag{11.28}$$

$$y_U = Y_U^* \left(1 + \alpha_Y \delta_U\right), \tag{11.29}$$

$$z_U = Z_U^* \left(1 + \alpha_Z \delta_U\right), \tag{11.30}$$

where δ_U is the mass fractionation factor for the sample. X_U^*, Y_U^*, Z_U^* are measured isotopic ratios and lower-case x, y and z are true isotopic ratios. α_X, α_Y and α_Z are the mass-difference coefficients between two isotopes. For Pb isotopic compositions, where $x = {}^{206}\text{Pb}/{}^{204}\text{Pb}$, $y = {}^{207}\text{Pb}/{}^{204}\text{Pb}$ and $z = {}^{208}\text{Pb}/{}^{204}\text{Pb}$, the mass-difference coefficients

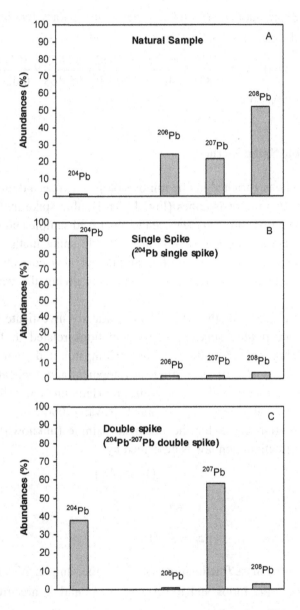

Fig. 11.3. Diagram comparing (A) the normal Pb sample, (B) a single ^{204}Pb spike and (C) a ^{204}Pb-^{207}Pb double spike.

are $\alpha_X = 206 - 204 = 2$, $\alpha_Y = 207 - 204 = 3$ and $\alpha_Z = 208 - 204 = 4$. And Eqs. (11.28), (11.29) and (11.30) become

$$\left({}^{206}Pb/{}^{204}Pb \right)_U^{true} = \left({}^{206}Pb/{}^{204}Pb \right)_U^{measured} \left[1 + 2\delta_U \right], \qquad (11.31)$$

$$\left({}^{207}Pb/{}^{204}Pb \right)_U^{true} = \left({}^{207}Pb/{}^{204}Pb \right)_U^{measured} \left[1 + 3\delta_U \right], \qquad (11.32)$$

$$\left({}^{208}Pb/{}^{204}Pb \right)_U^{true} = \left({}^{208}Pb/{}^{204}Pb \right)_U^{measured} \left[1 + 4\delta_U \right]. \qquad (11.33)$$

The fractionation curve for the mixture (spike+sample) from the linear mass fractionation law is described by

$$x_M = X_M^* \left(1 + \alpha_X \delta_M \right), \qquad (11.34)$$

$$y_M = Y_M^* \left(1 + \alpha_Y \delta_M \right), \qquad (11.35)$$

$$z_M = Z_M^* \left(1 + \alpha_Z \delta_M \right), \qquad (11.36)$$

where X_M^*, Y_M^* and Z_M^* are measured values for the sample-spike mixture and δ_M is the mass fractionation factor for the mixture. For Pb isotopes, Eqs. (11.34), (11.35) and (11.36) become

$$\left({}^{206}Pb/{}^{204}Pb \right)_M^{true} = \left({}^{206}Pb/{}^{204}Pb \right)_M^{measured} \left[1 + 2\delta_M \right], \qquad (11.37)$$

$$\left({}^{207}Pb/{}^{204}Pb \right)_M^{true} = \left({}^{207}Pb/{}^{204}Pb \right)_M^{measured} \left[1 + 3\delta_M \right], \qquad (11.38)$$

and

$$\left({}^{208}Pb/{}^{204}Pb \right)_M^{true} = \left({}^{208}Pb/{}^{204}Pb \right)_M^{measured} \left[1 + 4\delta_M \right]. \qquad (11.39)$$

Now we need the equation from the single spike as a mixing. For single spike, we only need to measure one isotope ratio. But for double spike, we need three isotopic ratios. According to Eq. (11.11), we have

$$x_M (1 + q) = x_N + q x_S, \qquad (11.40)$$

$$y_M (1 + q) = y_N + q y_S, \qquad (11.41)$$

$$z_M (1 + q) = z_N + q z_S, \qquad (11.42)$$

where

$$q = \frac{Ab_S^B}{Ab_N^B} \frac{S}{N},$$ (11.43)

and x_S, y_S and z_S are known for the spike. We have nine equations and nine unknowns in x_U, y_U, z_U, x_M, y_M, z_M, δ_U, δ_M and q. By eliminating six unknowns in x_U, y_U, z_U, x_M, y_M, z_M, the nine equations (11.34) to (11.42) may be reduced to the following three equations with three unknowns in δ_U, δ_M and q:

$$\left(X_M^* - X_S\right)q + (1+q)\alpha_X X_M^* \delta_M - \alpha_X X_U^* \delta_U = X_U^* - X_M^*,$$ (11.44)

$$\left(Y_M^* - Y_S\right)q + (1+q)\alpha_Y Y_M^* \delta_M - \alpha_X Y_U^* \delta_U = Y_U^* - Y_M^*,$$ (11.45)

$$\left(Z_M^* - Z_S\right)q + (1+q)\alpha_Z Z_M^* \delta_M - \alpha_X Z_U^* \delta_U = Z_U^* - Z_M^*.$$ (11.46)

We may write Eqs. (11.44), (11.45), (11.46) in a matrix form:

$$\begin{vmatrix} X_M^* - X_S & \alpha_X X_M^* & -\alpha_X X_U^* \\ Y_M^* - Y_S & \alpha_Y Y_M^* & -\alpha_Y Y_U^* \\ Z_M^* - Z_S & \alpha_Z Z_M^* & -\alpha_Z Z_U^* \end{vmatrix} \begin{vmatrix} q \\ (1+q)\delta_M \\ \delta_U \end{vmatrix} = \begin{vmatrix} X_U^* - X_M^* \\ Y_U^* - Y_M^* \\ Z_U^* - Z_M^* \end{vmatrix}.$$

(11.47)

The solution to Eq. (11.47) in a matrix form is (Galer, 1999)

$$\begin{bmatrix} q \\ (1+q)\delta_M \\ \delta_U \end{bmatrix} = \begin{vmatrix} X_M^* - X_S & \alpha_X X_M^* & -\alpha_X X_U^* \\ Y_M^* - Y_S & \alpha_Y Y_M^* & -\alpha_Y Y_U^* \\ Z_M^* - Z_S & \alpha_Z Z_M^* & -\alpha_Z Z_U^* \end{vmatrix}^{-1} \begin{vmatrix} X_U^* - X_M^* \\ Y_U^* - Y_M^* \\ Z_U^* - Z_M^* \end{vmatrix}.$$

(11.48)

Since the analytical solutions to Eq. (11.48) for δ_U, q, and $(1+q)\delta_M$ (and thus δ_M) have not been given before, here we provide in detail the analytical solutions as follows.

δ_U is obtained by

$$\delta_U = \frac{\alpha_X X_M^* A_1 + \alpha_Y Y_M^* A_2 + \alpha_Z Z_M^* A_3}{\Delta}, \quad (11.49)$$

where

$$A_1 = \left(Y_M^* - Y_S\right)\left(Z_U^* - Z_M^*\right) - \left(Y_U^* - Y_M^*\right)\left(Z_M^* - Z_S\right), \quad (11.50)$$

$$A_2 = \left(Z_M^* - Z_S\right)\left(X_U^* - X_M^*\right) - \left(Z_U^* - Z_M^*\right)\left(X_M^* - X_S\right), \quad (11.51)$$

$$A_3 = \left(X_M^* - X_S\right)\left(Y_U^* - Y_M^*\right) - \left(X_U^* - X_M^*\right)\left(Y_M^* - Y_S\right), \quad (11.52)$$

$$\begin{aligned}
\Delta = {} & \alpha_X \alpha_Z \left(Y_M^* - Y_S\right)\left(X_U^* Z_M^* - X_M^* Z_U^*\right) \\
& + \alpha_X \alpha_Y \left(Z_M^* - Z_S\right)\left(X_M^* Y_U^* - X_U^* Y_M^*\right) \\
& + \alpha_Y \alpha_Z \left(X_M^* - X_S\right)\left(Y_M^* Z_U^* - Y_U^* Z_M^*\right).
\end{aligned} \quad (11.53)$$

q is given by

$$q = \frac{\alpha_Y \alpha_Z B_1 + \alpha_X \alpha_Z B_2 + \alpha_X \alpha_Y B_3}{\Delta}, \quad (11.54)$$

where

$$B_1 = \left(X_U^* - X_M^*\right)\left(Y_M^* Z_U^* - Y_U^* Z_M^*\right), \quad (11.55)$$

$$B_2 = \left(Y_U^* - Y_M^*\right)\left(X_U^* Z_M^* - X_M^* Z_U^*\right), \quad (11.56)$$

$$B_3 = \left(Z_U^* - Z_M^*\right)\left(X_U^* Y_M^* - X_M^* Y_U^*\right). \quad (11.57)$$

And $(1+q)\delta_M$ is obtained by

$$(1+q)\delta_M = \frac{\alpha_X X_U^* C_1 + \alpha_Y Y_U^* C_2 + \alpha_Z Z_U^* C_3}{\Delta}, \quad (11.58)$$

where

$$C_1 = \left(Z_U^* - Z_M^*\right)\left(Y_M^* - Y_S\right) - \left(Y_U^* - Y_M^*\right)\left(Z_M^* - Z_S\right), \quad (11.59)$$

$$C_2 = \left(X_U^* - X_m^* \right)\left(Z_M^* - Z_S \right) - \left(Z_U^* - Z_M^* \right)\left(X_M^* - X_S \right), \quad (11.60)$$

$$C_3 = \left(Y_U^* - Y_M^* \right)\left(X_M^* - X_S \right) - \left(X_U^* - X_M^* \right)\left(Y_M^* - Y_S \right). \quad (11.61)$$

Note that δ_U, q and $(1+q)\delta_M$ have the same denominator Δ. It is noted that when $\alpha_X = 2$, $\alpha_Y = 3$, $\alpha_Z = 4$, which is the case for Pb isotopes, Eqs. (11.49), (11.54) and (11.58) collapse to the solutions given by Gale (1970). Also note that Johnson and Beard (1999) provided analytical solutions for a different matrix.

10.3. Two Double Spikes

The double spike method needs a run of un-spiked sample and the second run of the mixture. In contrast, two double spike method (Kuritani and Nakamura, 2003) requires two runs of mixtures spiked with different double spikes. Here we explain the theory of the [207]Pb-[204]Pb and [205]Pb-[204]Pb two double spikes (Kuritani and Nakamura, 2003). The true ratio of the first double spike mixture (e.g., x_{M1}) is related to the measured ratio of the first mixture (e.g., X_{M1}^*) by

$$x_{M1} = X_{M1}^* \left(1 + \alpha_X \delta_{M1} \right), \quad (11.62)$$

$$y_{M1} = Y_{M1}^* \left(1 + \alpha_Y \delta_{M1} \right), \quad (11.63)$$

$$z_{M1} = Z_{M1}^* \left(1 + \alpha_Z \delta_{M1} \right). \quad (11.64)$$

The mixing relationship between the true ratio of the first double spike mixture (e.g., x_{M1}) and the true ratio of the unknown (e.g., x_U) in the first mixture is

$$x_{M1}(1+p) = x_U + x_{S1}p, \quad (11.65)$$

$$y_{M1}(1+p) = y_U + y_{S1}p, \quad (11.66)$$

$$z_{M1}(1+p) = z_U + z_{S1}p. \quad (11.67)$$

The second double spike includes an artificially enriched isotope that does not exist in natural samples or in the first double spike. For example, [205]Pb in the [205]Pb-[204]Pb double spike does not exist in natural samples or in the first [207]Pb-[204]Pb double spike. Therefore, the second double spike has an additional isotope ratio w.

The true ratio for the second double spike mixture is related to the measured ratio of the second mixture by

$$w_{M2} = W_{M2}^* \left(1 + \alpha_W \delta_{M2} \right), \tag{11.68}$$

$$x_{M2} = X_{M2}^* \left(1 + \alpha_X \delta_{M2} \right), \tag{11.69}$$

$$y_{M2} = Y_{M2}^* \left(1 + \alpha_Y \delta_{M2} \right), \tag{11.70}$$

$$z_{M2} = Z_{M2}^* \left(1 + \alpha_Z \delta_{M2} \right). \tag{11.71}$$

The mixing relationship between the true ratio of the second mixture (e.g., x_{M2}) and the true ratio of the unknown (e.g., x_U) of the second mixture is

$$w_{M2}(1+q) = w_U + w_{S2}q, \tag{11.72}$$

$$x_{M2}(1+q) = x_U + x_{S2}q, \tag{11.73}$$

$$y_{M2}(1+q) = y_U + y_{S2}q, \tag{11.74}$$

$$z_{M2}(1+q) = z_U + z_{S2}q. \tag{11.75}$$

We can obtain seven equations: including three from the first mixture and four from the second mixture.

For the first mixture, from (11.62), (11.63), (11.64), (11.65), (11.66) and (11.67), we have

$$(X_{M1}^* - x_{S1})p + X_{M1}^*\alpha_X(1+p)\delta_{M1} - x_U = -X_{M1}^*, \tag{11.76}$$

$$(Y_{M1}^* - x_{S1})p + Y_{M1}^*\alpha_X(1+p)\delta_{M1} - y_U = -Y_{M1}^*, \tag{11.77}$$

$$(Z_{M1}^* - z_S)p + Z_{M1}^*\alpha_X(1+p)\delta_{M1} - z_U = -Z_{M1}^*. \tag{11.78}$$

For the second mixture, from (11.68), (11.69), (11.70), (11.71), (11.72), (11.73), (11.74) and (11.75), we get

$$(W_{M2}^* - w_S)q + W_{M2}^*\alpha_X(1+p)\delta_{M2} - w_U = -W_{M2}^*, \qquad (11.79)$$

$$(X_{M2}^* - x_{S2})q + X_{M2}^*\alpha_X(1+p)\delta_{M2} - x_U = -X_{M2}^*, \qquad (11.80)$$

$$(Y_{M2}^* - y_S)q + Y_{M2}^*\alpha_X(1+p)\delta_{M2} - y_U = -Y_{M2}^*, \qquad (11.81)$$

$$(Z_{M2}^* - z_S)q + Z_{M2}^*\alpha_X(1+p)\delta_{M2} - z_U = -Z_{M2}^*. \qquad (11.82)$$

We have seven unknowns in p, q, x_U, y_U, z_U, δ_{M1}, δ_{M2} in seven linear equations (11.76) to (11.82) and we can obtain unique solutions. Let $r_{M1} = (1+p)\delta_{M1}$ and $r_{M2} = (1+q)\delta_{M2}$, we can write Eqs. (11.76) to (11.82) in a matrix form

$$\begin{vmatrix} X_{M1}^* - x_{S1} & 0 & \alpha_X X_{M1}^* & 0 & -1 & 0 & 0 \\ Y_{M1}^* - y_{S1} & 0 & \alpha_Y Y_{M1}^* & 0 & 0 & -1 & 0 \\ Z_{M1}^* - z_{S1} & 0 & \alpha_Z Z_{M1}^* & 0 & 0 & 0 & -1 \\ 0 & W_{M2}^* - w_{S2} & 0 & \alpha_W W_{M2}^* & 0 & 0 & 0 \\ 0 & X_{M2}^* - x_{S2} & 0 & \alpha_X X_{M2}^* & -1 & 0 & 0 \\ 0 & Y_{M2}^* - y_{S2} & 0 & \alpha_Y Y_{M2}^* & 0 & -1 & 0 \\ 0 & Z_{M2}^* - z_{S2} & 0 & \alpha_Z Z_{M2}^* & 0 & 0 & -1 \end{vmatrix} \begin{vmatrix} p \\ q \\ r_{M1} \\ r_{M2} \\ x_U \\ y_U \\ z_U \end{vmatrix} = \begin{vmatrix} -X_{M1}^* \\ -Y_{M1}^* \\ -Z_{M1}^* \\ -W_{M2}^* \\ -X_{M2}^* \\ -Y_{M2}^* \\ -Z_{M2}^* \end{vmatrix}$$

$$(11.83)$$

The solution to Eq. (11.83) in the matrix form is (Kuritani and Nakamura, 2003)

$$\begin{vmatrix} p \\ q \\ r_{M1} \\ r_{M2} \\ x_U \\ y_U \\ z_U \end{vmatrix} = \begin{vmatrix} X_{M1}^* - x_{S1} & 0 & \alpha_X X_{M1}^* & 0 & -1 & 0 & 0 \\ Y_{M1}^* - y_{S1} & 0 & \alpha_Y Y_{M1}^* & 0 & 0 & -1 & 0 \\ Z_{M1}^* - z_{S1} & 0 & \alpha_Z Z_{M1}^* & 0 & 0 & 0 & -1 \\ 0 & W_{M2}^* - w_{S2} & 0 & \alpha_W W_{M2}^* & 0 & 0 & 0 \\ 0 & X_{M2}^* - x_{S2} & 0 & \alpha_X X_{M2}^* & -1 & 0 & 0 \\ 0 & Y_{M2}^* - y_{S2} & 0 & \alpha_Y Y_{M2}^* & 0 & -1 & 0 \\ 0 & Z_{M2}^* - z_{S2} & 0 & \alpha_Z Z_{M2}^* & 0 & 0 & -1 \end{vmatrix}^{-1} \begin{vmatrix} -X_{M1}^* \\ -Y_{M1}^* \\ -Z_{M1}^* \\ -W_{M2}^* \\ -X_{M2}^* \\ -Y_{M2}^* \\ -Z_{M2}^* \end{vmatrix}$$

$$(11.84)$$

Note that $w_U = 0$ in (11.79) (for example, we use $^{205}\text{Pb}/^{204}\text{Pb}$ as w_U and $w_U = 0$ for natural samples) because natural samples do not have short-lived ^{205}Pb. Otherwise, if $w_U \neq 0$, we have eight unknowns in seven equations and we can not obtain unique solutions.

10.4. Summary of Equations

1) Single spike method

The concentration in the normal sample can be obtained by

$$C_N = C_{Spike} \frac{M_S}{M_N} \frac{W_N}{W_S} \left[\frac{Ab_S^A - R_m Ab_S^B}{R_m Ab_N^B - Ab_N^A} \right],$$

or, equivalently, by

$$C_N = C_{Spike} \frac{R_S - R_M}{R_M - R_N} \frac{M_S}{M_N} \frac{W_N}{W_S} \frac{Ab_S^B}{Ab_N^B}.$$

The relationship between the isotopic ratio in the mixture, normal sample and spike is

$$R_M (1 + q) = R_N + q R_S,$$

$$\text{where } q = \frac{Ab_S^B S}{Ab_N^B N}.$$

Magnification factor:

$$Q = \frac{R_M (R_S - R_N)}{(R_M - R_N)(R_S - R_M)}.$$

Optimal R_M:

$$R_M = \sqrt{R_N R_S}.$$

2) Double spike method

$$\delta_U = \frac{\alpha_X X_M^* A_1 + \alpha_Y Y_M^* A_2 + \alpha_Z Z_M^* A_3}{\Delta},$$

$$q = \frac{\alpha_Y \alpha_Z B_1 + \alpha_X \alpha_Z B_2 + \alpha_X \alpha_Y B_3}{\Delta},$$

$$(1+q)\delta_M = \frac{\alpha_X X_U^* C_1 + \alpha_Y Y_U^* C_2 + \alpha_Z Z_U^* C_3}{\Delta},$$

where

$$\Delta = \alpha_X \alpha_Z \left(Y_M^* - Y_S \right) \left(X_U^* Z_M^* - X_M^* Z_U^* \right)$$
$$+ \alpha_X \alpha_Y \left(Z_M^* - Z_S \right) \left(X_M^* Y_U^* - X_U^* Y_M^* \right)$$
$$+ \alpha_Y \alpha_Z \left(X_M^* - X_S \right) \left(Y_M^* Z_U^* - Y_U^* Z_M^* \right),$$

and A_1, A_2, A_3, B_1, B_2, B_3, C_1, C_2, C_3, are given by (11.50), (11.51), (11.52), (11.55), (11.56), (11.57), (11.59), (11.60) and (11.61), respectively.

References

Compston, W. and Oversby, V.M., 1969. Lead isotope analysis using a double spike. J. Geophys. Res., 74: 4338-4348.

Dodson, M.H., 1963. A theoretical study of the use of internal standards for precise isotopic analysis by the surface ionization techniques: Part I--general first-order algebraic solutions. J. Sci. Instrum., 40: 289-295.

Faure, G., 1986. Principles of Isotope Geology. John Wiley and Sons, Inc., New York, 589 pp.

Gale, N.H., 1970. A solution in closed form for lead isotopic analysis using a double spike. Chem. Geol., 6: 305-310.

Galer, S.J.G., 1999. Optimal double and triple spiking for high precision lead isotopic measurement. Chem. Geol., 157: 255-274.

Hamelin, B., Manhes, G., Albarede, F. and Allegre, C.J., 1985. Precise lead isotope measurements by the double spike technique. Geochim. Cosmochim. Acta, 49: 173-182.

Johnson, C.M. and Beard, B.L., 1999. Correction of instrumentally produced mass fractionation during isotopic analysis of Fe by thermal ionization mass spectrometry. International J. Mass Spectrometry, 193: 87-99.

Kuritani, T. and Nakamura, E., 2003. Highly precise and accurate isotopic analysis of small amounts of Pb using ^{205}Pb-^{204}Pb and ^{207}Pb-^{204}Pb, two double spikes. J. Ana. At. Spectrom., 18: 1464-1470.

Thirlwall, M.F., 2000. Inter-labortory and other errors in Pb isotope analyses using a ^{207}Pb-^{204}Pb double spike. Chem. Geol., 163: 299-322.

Chapter 12

Pb Isotope Modeling

Pb isotope evolution is complicated. Mathematical analysis of the Pb isotope diagrams can help understand challenging problems related to Pb isotope evolution. Here we conduct new theoretical analyses of the fundamental equations related to four useful diagrams in Pb isotope geochemistry: the Tera-Wasserburg diagram, the conventional U-Pb Concordia diagram, the Holmes-Houtermans common Pb evolution model and the two-stage Pb evolution model. The U-Th-Pa Concordia diagram will also be treated in a similar manner. You may find that these quantitative treatments are quite stimulating. At the end of this chapter, we talk about the error analyses of U-Pb isotope data.

12.1. Why is the Tera-Wasserburg Concordia Diagram Concave Upward?

In the Tera-Wasserburg Concordia (Tera and Wasserburg, 1972), $^{238}U/^{206}Pb$ ratios are plotted directly against the $^{207}Pb/^{206}Pb$ ratio. The equations for $^{238}U/^{206}Pb$ and $^{207}Pb/^{206}Pb$ are given by:

$$\frac{^{238}U}{^{206}Pb} = \frac{1}{e^{\lambda_{238}t} - 1},$$ (12.1)

$$\frac{^{207}Pb}{^{206}Pb} = \frac{^{235}U}{^{238}U}\frac{e^{\lambda_{235}t} - 1}{e^{\lambda_{238}t} - 1},$$ (12.2)

where $\lambda_{238} = 1.55125 \times 10^{-10}$ y^{-1} and $\lambda_{235} = 9.8485 \times 10^{-10}$ y^{-1} are decay constants for ^{238}U and ^{235}U, respectively.

248

Because the second derivative (d^2y/dx^2) of a function $(y = f(x))$ determines whether the curve of the function is concave upward or down, to answer the question of this section, we need to derive the second derivatives of the Tera-Wasserburg Concordia plot.

For the Tera-Wasserburg Concordia diagram (Fig. 12.1), we have

$$x = {}^{238}U \big/ {}^{206}Pb,$$

$$y = {}^{207}Pb \big/ {}^{206}Pb.$$

Differentiation of x in Eq. (12.1) with respect to t gives

$$\frac{\partial x}{\partial t} = \frac{-\lambda_{238} e^{\lambda_{238}t}}{\left(e^{\lambda_{238}t} - 1\right)^2}. \tag{12.3}$$

Differentiation of y in Eq. (12.2) with respect to t yields

$$\frac{\partial y}{\partial t} = \frac{{}^{235}U}{{}^{238}U} \frac{\lambda_{235} e^{\lambda_{235}t}\left(e^{\lambda_{238}t} - 1\right) - \lambda_{238} e^{\lambda_{238}t}\left(e^{\lambda_{235}t} - 1\right)}{\left(e^{\lambda_{238}t} - 1\right)^2}$$

$$= \frac{{}^{235}U \big/ {}^{238}U}{\left(e^{\lambda_{238}t} - 1\right)^2}\left[\lambda_{238} e^{\lambda_{238}t} - \lambda_{235} e^{\lambda_{235}t} + \left(\lambda_{235} - \lambda_{238}\right) e^{(\lambda_{235} + \lambda_{238})t}\right].$$

$$\tag{12.4}$$

From Eqs. (12.3) and (12.4) we obtain the first derivative of y with respect to x by using chain rule

$$\frac{\partial y}{\partial x} = \frac{\partial y}{\partial t} \bigg/ \left(\frac{\partial x}{\partial t}\right)$$

$$= -\frac{{}^{235}U \big/ {}^{238}U}{\lambda_{238} e^{\lambda_{238}t}}\left[\begin{array}{l}\lambda_{238} e^{\lambda_{238}t} - \lambda_{235} e^{\lambda_{235}t} \\ + \left(\lambda_{235} - \lambda_{238}\right) e^{(\lambda_{235} + \lambda_{238})t}\end{array}\right] \tag{12.5}$$

$$= \frac{{}^{235}U}{{}^{238}U}\left\{\frac{\lambda_{235}}{\lambda_{238}}\left[e^{(\lambda_{235} - \lambda_{238})t} - e^{\lambda_{235}t}\right] + e^{\lambda_{235}t} - 1\right\}.$$

Differentiation of the above $\partial y / \partial x$ in (12.5) with respect to t gives

$$\frac{\partial(\partial y/\partial x)}{\partial t} = \frac{^{235}\text{U}}{^{238}\text{U}}\left[\begin{array}{l}\dfrac{\lambda_{235}}{\lambda_{238}}(\lambda_{235}-\lambda_{238})e^{(\lambda_{235}-\lambda_{238})t}\\[2mm]+\lambda_{235}e^{\lambda_{235}t}-\dfrac{\lambda_{235}^{\ 2}}{\lambda_{238}}e^{\lambda_{235}t}\end{array}\right] \qquad (12.6)$$

$$= \frac{^{235}\text{U}}{^{238}\text{U}}\frac{\lambda_{235}}{\lambda_{238}}(\lambda_{235}-\lambda_{238})\left[e^{(\lambda_{235}-\lambda_{238})t}-e^{\lambda_{235}t}\right].$$

The key to obtain $\partial^2 y/\partial x^2$ is to use the following chain rule:

$$\frac{\partial^2 y}{\partial x^2} = \frac{\partial(\partial y/\partial x)}{\partial x} = \frac{\partial(\partial y/\partial x)}{\partial t}\frac{\partial t}{\partial x}$$

$$= \frac{\partial(\partial y/\partial x)}{\partial t}\bigg/\left(\frac{\partial x}{\partial t}\right). \qquad (12.7)$$

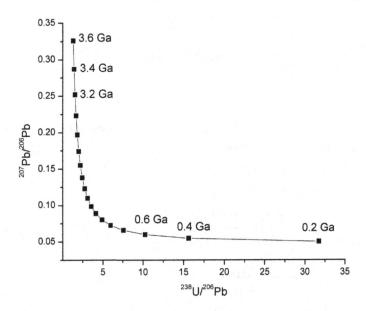

Fig. 12.1. The Tera-Wasserburg $^{207}\text{Pb}/^{206}\text{Pb}$ vs. $^{238}\text{U}/^{206}\text{Pb}$ plot from Eqs. (12.1) and (12.2). Ga=10^9 years.

From (12.3), (12.6), (12.7), we obtain the second derivative of y with respect to x as

$$\frac{\partial^2 y}{\partial x^2} = \frac{\partial(\partial y/\partial x)}{\partial t} \bigg/ \left(\frac{\partial x}{\partial t}\right)$$

$$= \frac{^{235}U}{^{238}U}\frac{\lambda_{235}}{\lambda_{238}}(\lambda_{235}-\lambda_{238})\left[e^{(\lambda_{235}-\lambda_{238})t}-e^{\lambda_{235}t}\right] \bigg/ \left[\frac{-\lambda_{238}e^{\lambda_{238}t}}{(e^{\lambda_{238}t}-1)^2}\right]$$

$$= \frac{^{235}U}{^{238}U}\frac{\lambda_{235}}{\lambda_{238}^2}(\lambda_{235}-\lambda_{238})e^{-\lambda_{238}t}(e^{\lambda_{238}t}-1)^2\left[e^{\lambda_{235}t}-e^{(\lambda_{235}-\lambda_{238})t}\right].$$

$$(12.8)$$

Since

$$\lambda_{235}-\lambda_{238}>0,$$

$$e^{\lambda_{235}t}-e^{(\lambda_{235}-\lambda_{238})t}>0,$$

and all other terms on the right-hand side in Eq. (12.8) are positive, the second derivative $\partial^2 y/\partial x^2 > 0$ everywhere. Therefore, the curves in the Tera-Wasserburg Concordia diagram are concave up everywhere (Fig. 12.1).

12.2. Why is the Conventional Concordia Plot Concave Down?

The conventional Concordia diagram (Wetherill, 1956) is the $^{206}Pb^*/^{238}U$ vs. $^{207}Pb^*/^{235}U$ plot, where * denotes the radiogenic portion of Pb isotopes. This diagram also finds wide applications in zircon U-Pb geochronology. For U-Pb zircon dating, the daughter-parent ratios are given by

$$^{206}Pb^*/^{238}U = e^{\lambda_{238}t}-1, \qquad (12.9)$$

$$^{207}Pb^*/^{235}U = e^{\lambda_{235}t}-1. \qquad (12.10)$$

In conventional Concordia diagram (Fig. 12.2), $x = {}^{207}Pb^*/^{235}U$ and $y = {}^{206}Pb^*/^{238}U$.

The first derivative of y with respect to time t from Eq. (12.9) is

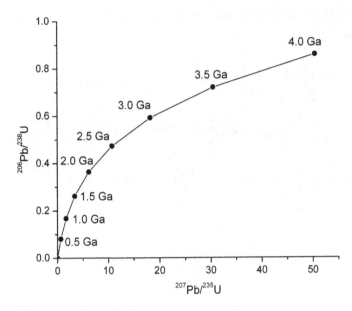

Fig. 12.2. The $^{206}Pb/^{238}U$ vs. $^{207}Pb/^{235}U$ Concordia plot according to Eqs. (12.9) and (12.10).

$$\frac{\partial y}{\partial t} = \lambda_{238} e^{\lambda_{238} t} . \qquad (12.11)$$

Similarly, the first derivative of x with respect to t from Eq. (12.10) is

$$\frac{\partial x}{\partial t} = \lambda_{235} e^{\lambda_{235} t} . \qquad (12.12)$$

From Eqs. (12.11) and (12.12), we obtain the first derivative of y with respect to x by using the chain rule

$$\frac{\partial y}{\partial x} = \frac{\partial y}{\partial t} \Big/ \left(\frac{\partial x}{\partial t} \right) = \frac{\lambda_{238}}{\lambda_{235}} e^{(\lambda_{238} - \lambda_{235}) t} . \qquad (12.13)$$

All terms on the right hand side of Eq. (12.13) is positive and thus $\partial y/\partial x > 0$ everywhere. It is not surprising that a positive relationship between $^{206}Pb/^{238}U$ and $^{207}Pb/^{235}U$ is displayed in the conventional Concordia plot (Fig. 12.2).

To obtain the second derivative of y with respect to x, or $\partial^2 y / \partial x^2$, we need to first obtain the derivative of $\partial y / \partial x$ with respect to t from Eq. (12.13):

$$\frac{\partial (\partial y / \partial x)}{\partial t} = \frac{\lambda_{238}(\lambda_{238} - \lambda_{235})}{\lambda_{235}} e^{(\lambda_{238} - \lambda_{235})t}. \qquad (12.14)$$

From Eqs. (12.12) and (12.14) we find

$$\frac{\partial^2 y}{\partial x^2} = \frac{\partial (\partial y / \partial x)}{\partial t} \bigg/ \left(\frac{\partial x}{\partial t}\right)$$

$$= \frac{\lambda_{238}(\lambda_{238} - \lambda_{235})}{\lambda_{235}^2} e^{(\lambda_{238} - 2\lambda_{235})t}. \qquad (12.15)$$

Since $\lambda_{238} - \lambda_{235} < 0$ in Eq. (12.15), the second derivative $\partial^2 y / \partial x^2 < 0$ everywhere. Therefore, the conventional Concordia plot is concave down everywhere (Fig. 12.2).

12.3. Why is the Holmes-Houtermans Model Concave Down?

For the Tera-Wasserburg plot and conventional Concordia plot, we measure a mineral (e.g., zircon) that is enriched in U and initially contains little unradiogenic (common) Pb.

For Holmes-Houtermans model (Holmes, 1946; Houtermans, 1946), we chose a mineral (e.g., galena, PbS) that contains significant amount of Pb but very little U. Let t be the formation age of the galena. In the Holmes-Houtermans model of lead evolution, Pb evolved between 4.57 Ga (T) and time (t) in a reservoir having uniform $^{238}U/^{204}Pb$ (μ) of 7.192. Galena was formed and separated from the reservoir at time t. From t to the Present, there is no radiogenic Pb growth in the galena, or, Pb isotopic compositions were fixed between t and present. The equations for Pb isotope compositions in the Holmes-Houtermans model are given by

$$\left(\frac{^{206}Pb}{^{204}Pb}\right)_t = \left(\frac{^{206}Pb}{^{204}Pb}\right)_T + \frac{^{238}U}{^{204}Pb}\left(e^{\lambda_{238}T} - e^{\lambda_{238}t}\right), \qquad (12.16)$$

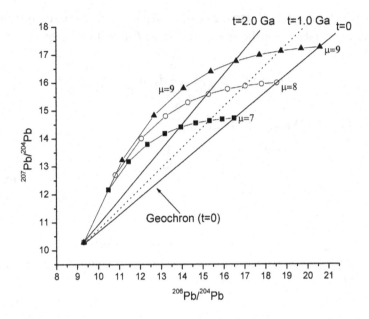

Fig. 12.3. The one-stage Pb evolution model of Holmes-Houtermans according to Eqs. (12.16) and (12.17). The straight lines are isochrones for different values of t. The curved lines are Pb growth curves for U-Pb systems having present-day μ ($^{238}U/^{204}Pb$) values of 7 (solid square), 8 (open circle), and 9 (solid triangle), respectively. Present-day $^{238}U/^{235}U=137.88$.

$$\left(\frac{^{207}Pb}{^{204}Pb}\right)_t = \left(\frac{^{207}Pb}{^{204}Pb}\right)_T + \frac{^{235}U}{^{204}Pb}\left(e^{\lambda_{235}T} - e^{\lambda_{235}t}\right). \qquad (12.17)$$

where $\left(\dfrac{^{206}Pb}{^{204}Pb}\right)_t$ and $\left(\dfrac{^{207}Pb}{^{204}Pb}\right)_t$ are the Pb isotopic compositions at time t to form mineral galena and $\left(\dfrac{^{206}Pb}{^{204}Pb}\right)_T$ and $\left(\dfrac{^{207}Pb}{^{204}Pb}\right)_T$ are the Pb isotopic ratios when the Earth was formed.
Let

$$x = \left(\frac{^{206}Pb}{^{204}Pb}\right)_t,$$

$$y = \left(\frac{^{207}Pb}{^{204}Pb} \right)_t.$$

Differentiation of x in Eq. (12.16) with respect to t gives

$$\frac{\partial x}{\partial t} = -\frac{^{238}U}{^{204}Pb} \lambda_{238} e^{\lambda_{238}t} . \qquad (12.18)$$

Differentiation of y in Eq. (12.17) with respect to t yields

$$\frac{\partial y}{\partial t} = -\frac{^{235}U}{^{204}Pb} \lambda_{235} e^{\lambda_{235}t} . \qquad (12.19)$$

By dividing Eq. (12.19) over Eq. (12.18), we obtain the first derivative of y with respective to x

$$\frac{\partial y}{\partial x} = \frac{^{235}U}{^{238}U} \frac{\lambda_{235}}{\lambda_{238}} e^{(\lambda_{235} - \lambda_{238})t} . \qquad (12.20)$$

Note that, according to Eq. (12.20), the slope ($\partial y/\partial x$) of the line for a fixed t is independent of $^{238}U/^{204}Pb$ ratio, or μ, which is consistent with the straight lines for fixed t in Fig. 12.3.

Differentiation of $\partial y/\partial x$ from Eq. (12.20) with respect to t yields

$$\frac{\partial(\partial y/\partial x)}{\partial t} = \frac{^{235}U}{^{238}U} \frac{\lambda_{235}}{\lambda_{238}} (\lambda_{235} - \lambda_{238}) e^{(\lambda_{235} - \lambda_{238})t} . \qquad (12.21)$$

From Eqs. (12.18) and (12.21) we obtain

$$\frac{\partial^2 y}{\partial x^2} = \frac{\partial(\partial y/\partial x)}{\partial t} \bigg/ \left(\frac{\partial x}{\partial t} \right)$$

$$= -\frac{1}{\left(^{238}U/^{204}Pb \right)} \frac{^{235}U}{^{238}U} \frac{\lambda_{235}}{\lambda_{238}^2} (\lambda_{235} - \lambda_{238}) e^{(\lambda_{235} - 2\lambda_{238})t} \qquad (12.22)$$

$$< 0.$$

Since the second derivative is negative everywhere, the plot is concave down everywhere.

It should be noted that, although the first derivative is independent of $^{208}U/^{204}Pb$, or μ, the second derivative depends on μ, resulting in different extents of curvatures. According to Eq. (12.22), the higher the μ, the lower the absolute value of $\partial^2 y/\partial x^2$, and thus smaller the extents of the curvature. This analysis is consistent with Fig. 12.3 in that the curve for $\mu = 9$ displays smaller extent of curvature than that for $\mu = 7$.

12.4. Why is the Two-stage Model of Pb Evolution also Concave Down?

In the Stacey-Kremers two-stage model of lead evolution (Stacey and Kramers, 1975), Pb evolved between 4.57 Ga (T) and 3.7 Ga (t_1) in a reservoir having uniform $^{238}U/^{204}Pb$ (μ_1) of 7.192. At a time of t_1, the value of μ was changed by geochemical differentiation to $\mu_2 = 9.735$. At a later time t_2, Pb was separated from the second-stage reservoir. The Pb isotopic compositions of such separated lead in minerals like galena have remained unchanged since t_2 till the Present. T and t_1 are fixed but t_2 is the unknown.

The equations for the two-stage model of Pb evolution are

$$\left(\frac{^{206}Pb}{^{204}Pb}\right)_t = \left(\frac{^{206}Pb}{^{204}Pb}\right)_T + \mu_1\left(e^{\lambda_{238}T} - e^{\lambda_{238}t_1}\right) + \mu_2\left(e^{\lambda_{238}t_1} - e^{\lambda_{238}t_2}\right), \qquad (12.23)$$

$$\left(\frac{^{207}Pb}{^{204}Pb}\right)_t = \left(\frac{^{207}Pb}{^{204}Pb}\right)_T + \frac{\mu_1}{137.88}\left(e^{\lambda_{235}T} - e^{\lambda_{235}t_1}\right)$$
$$+ \frac{\mu_2}{137.88}\left(e^{\lambda_{235}t_1} - e^{\lambda_{235}t_2}\right), \qquad (12.24)$$

where $\left(\dfrac{^{206}Pb}{^{204}Pb}\right)_T$ and $\left(\dfrac{^{207}Pb}{^{204}Pb}\right)_T$ are the primordial values when the Earth was formed and are known. 137.88 is the ratio $^{238}U/^{235}U$. $\mu_1 = (^{238}U/^{204}Pb)_1 = 7.192$ and $\mu_1 = (^{238}U/^{204}Pb)_2 = 9.735$

Let

$$x = \left(\frac{^{206}Pb}{^{204}Pb}\right)_t,$$

$$y = \left(\frac{^{207}Pb}{^{204}Pb}\right)_t.$$

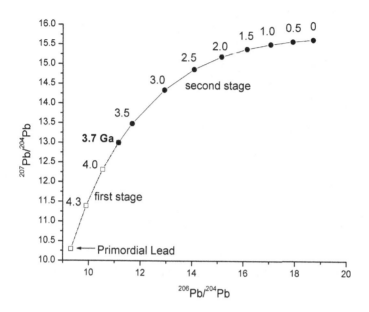

Fig. 12.4. The two-stage Pb evolution model of Stacey and Kramers (1975) from Eqs. (12.23) and (12.24). $\mu = 7.192$ for the first stage (4.57 Ga to 3.7 Ga) and $\mu = 9.735$ for the second stage (3.7 Ga to present).

Differentiation of Eq. (12.23) with respect to t_2 gives

$$\frac{\partial x}{\partial t_2} = -\mu_2 \lambda_{238} e^{\lambda_{238} t_2} . \tag{12.25}$$

Differentiation of Eq. (12.24) with respect to t_2 yields

$$\frac{\partial y}{\partial t_2} = -\frac{\mu_2}{137.88} \lambda_{235} e^{\lambda_{235} t} . \tag{12.26}$$

By using the chain rule, we obtain from Eqs. (12.25) and (12.26)

$$\frac{\partial y}{\partial x} = \frac{1}{137.88} \frac{\lambda_{235}}{\lambda_{238}} e^{(\lambda_{235} - \lambda_{238}) t_2} . \tag{12.27}$$

Note that, according to Eq. (12.27), the first derivative dy/dx is independent of μ_2.

The derivative of $\partial y/\partial x$ with respect to t_2 is

$$\frac{\partial(\partial y/\partial x)}{\partial t_2} = \frac{^{235}U}{^{238}U} \frac{\lambda_{235}}{\lambda_{238}} (\lambda_{235} - \lambda_{238}) e^{(\lambda_{235}-\lambda_{238})t_2} . \qquad (12.28)$$

From Eqs. (12.25) and (12.28) we finally find

$$\frac{\partial^2 y}{\partial x^2} = \frac{\partial(\partial y/\partial x)}{\partial t_2} \bigg/ \left(\frac{\partial x}{\partial t_2} \right)$$

$$= -\frac{1}{\mu_2} \frac{^{235}U}{^{238}U} \frac{\lambda_{235}}{\lambda_{238}^2} (\lambda_{235} - \lambda_{238}) e^{(\lambda_{235}-2\lambda_{238})t_2} . \qquad (12.29)$$

Since $\partial^2 y/\partial x^2 < 0$ everywhere, the curves in the two-stage Pb evolution model are concave down everywhere. According to Eq. (12.29), the extent of the curvature during the second stage is dependent on μ_2.

12.5. Why is the U-Th-Pa Concordia Diagram Concave Down Everywhere?

The Concordia U-Th-Pa diagram (Cheng et al., 1998) is $(^{231}Pa/^{235}U)$ vs. $(^{230}Th/^{238}U)$ plot with the following equations for unsupported ^{230}Th and ^{231}Pb (Ku, 1968; Ivanovich and Harmon, 1992):

$$(^{230}Th/^{238}U) = 1 - e^{-\lambda_{230}t} ,$$

$$(^{231}Pa/^{235}U) = 1 - e^{-\lambda_{231}t} .$$

In the Concordia U-Th-Pa diagram, we have

$$x = (^{230}Th/^{238}U) = 1 - e^{-\lambda_{230}t} , \qquad (12.30)$$

and

$$y = (^{231}Pa/^{235}U) = 1 - e^{-\lambda_{231}t} . \qquad (12.31)$$

The first derivative of y from Eq. (12.31) with respect to time t is

$$\frac{\partial y}{\partial t} = \lambda_{231} e^{-\lambda_{231}t} . \qquad (12.32)$$

Similarly, the first derivative of x from Eq. (12.30) with respect to t is

$$\frac{\partial x}{\partial t} = \lambda_{230} e^{-\lambda_{230}t} . \qquad (12.33)$$

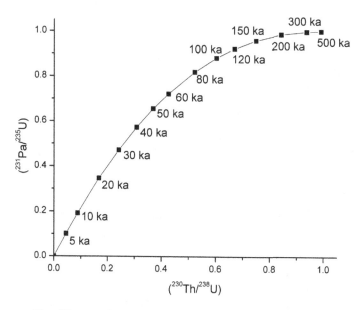

Fig. 12.5. $(^{231}\text{Pa}/^{235}\text{U})$ vs. $(^{230}\text{Th}/^{238}\text{U})$ Concordia diagram according to Eqs. (12.30) and (12.31). ka $= 1,000$ years .

From Eqs. (12.32) and (12.33), by using the chain rule, we obtain the first derivative of y with respect to x

$$\frac{\partial y}{\partial x} = \frac{\partial y}{\partial t} \bigg/ \left(\frac{\partial x}{\partial t}\right) = \frac{\lambda_{231}}{\lambda_{230}} e^{(\lambda_{230} - \lambda_{231})t} . \tag{12.34}$$

All terms on the right hand side of Eq. (12.34) is positive and thus $\partial y/\partial x > 0$ everywhere, as displayed by the positive relationship between $(^{230}\text{Th}/^{238}\text{U})$ and $(^{231}\text{Pa}/^{235}\text{U})$ in the U-Th-Pa Concordia plot (Fig. 12.5). To obtain the second derivative of y with respect to x, or $\partial^2 y/\partial x^2$, we need to first obtain the derivative of $\partial y/\partial x$ with respect to t from Eq. (12.34):

$$\frac{\partial(\partial y/\partial x)}{\partial t} = \frac{\lambda_{231}(\lambda_{230} - \lambda_{231})}{\lambda_{230}} e^{-(\lambda_{230} - \lambda_{231})t} . \tag{12.35}$$

By using chain rule and by combining Eqs. (12.33) and (12.35), we obtain

$$\frac{\partial^2 y}{\partial x^2} = \frac{\partial (\partial y/\partial x)}{\partial t} \bigg/ \left(\frac{\partial x}{\partial t}\right)$$

$$= \frac{\lambda_{231}(\lambda_{230} - \lambda_{231})}{\lambda_{230}^2} e^{(2\lambda_{230} - \lambda_{231})t}. \tag{12.36}$$

Since $\lambda_{230} = 9.217 \times 10^{-6}$ y^{-1} and $\lambda_{231} = 2.134 \times 10^{-5}$ y^{-1}, we have $\lambda_{230} - \lambda_{231} < 0$. The second derivative $\partial^2 y/\partial x^2 < 0$ everywhere in Eq. (12.36). Therefore, the U-Th-Pa Concordia plot is concave down everywhere.

12.6. Some Calculations Related to Pb Isotope Geochemistry

12.6.1. What is Concordia age?

We can calculate two independent ages based on ^{206}Pb/^{238}U and ^{207}Pb/^{235}U:

$$x = ^{207}Pb/^{235}U = e^{\lambda_{235}t} - 1, \tag{12.37}$$

$$y = ^{206}Pb/^{238}U = e^{\lambda_{238}t} - 1. \tag{12.38}$$

If the measured ^{207}Pb/^{235}U and ^{206}Pb/^{238}U plot on the Concordia curve (Fig. 12.2), then these two ages are consistent. In this case, we may calculate the Concordia age that is the best estimate of a single assumed Concordia point. The Concordia age is essentially a mathematically adjusted age.

A general goal is to minimize the sums of the squares of the N error weighted residuals (Titterington and Halliday, 1979; Ludwig, 1998)

$$S = \sum_{1}^{N} v_i^T \Omega_i v_i. \tag{12.39}$$

For a single assumed Concordia point ($(N = 1)$, Eq. (12.39) reduces to

$$S = v^T \Omega v, \tag{12.40}$$

where T denotes transpose, v is the vector of the residue of the point

$$v = \begin{vmatrix} X - x \\ Y - y \end{vmatrix} = \begin{vmatrix} R \\ r \end{vmatrix}, \tag{12.41}$$

and Ω is the variance-covariance matrix for the residuals

$$\Omega = \begin{vmatrix} \Omega_{11} & \Omega_{12} \\ \Omega_{21} & \Omega_{22} \end{vmatrix} = \begin{vmatrix} \sigma_x^2 & \text{cov}(x, y) \\ \text{cov}(x, y) & \sigma_y^2 \end{vmatrix}, \tag{12.42}$$

and X and Y are the measured Pb isotope values and x and y are the true values.

Substitution of Eqs. (12.37) and (12.38) into (12.41) results in

$$v = \begin{vmatrix} X - e^{\lambda_{235}t} + 1 \\ Y - e^{\lambda_{238}t} + 1 \end{vmatrix} = \begin{vmatrix} R \\ r \end{vmatrix}, \tag{12.43}$$

where t is the true Concordia age.

Substitution of (12.41) and (12.43) into Eq. (12.40) yields

$$S = \begin{vmatrix} R & r \end{vmatrix} \begin{vmatrix} \Omega_{11} & \Omega_{12} \\ \Omega_{21} & \Omega_{22} \end{vmatrix} \begin{vmatrix} R \\ r \end{vmatrix}$$

$$= \begin{vmatrix} R\Omega_{11} + r\Omega_{21} & R\Omega_{12} + r\Omega_{22} \end{vmatrix} \begin{vmatrix} R \\ r \end{vmatrix} \tag{12.44}$$

$$= R^2\Omega_{11} + Rr(\Omega_{12} + \Omega_{21}) + r^2\Omega_{22}$$

$$= R^2\Omega_{11} + 2Rr\Omega_{12} + r^2\Omega_{22}.$$

Since $R = X - e^{\lambda_{235}t} + 1$ and $r = Y - e^{\lambda_{238}t} + 1$, there is only one unknown t in Eq. (12.44). By setting $dS/dt = 0$ and solving it numerically, we can obtain the Concordia age.

12.6.2. *Error propagation related to Pb isotopes*

In this session, we show three examples concerning the error propagation of U-Pb isotope data. More detailed accounts of the error analysis of U-Pb isotope data have been documented in literature (Ludwig, 1980; Ludwig, 1998; Ludwig, 2003).

Example. Estimate the error of $^{206}Pb/^{204}Pb$ from measured $^{206}Pb/^{204}Pb$ and mass fractionation factor, given the measured $^{206}Pb/^{204}Pb=$ 18.792 ± 0.005 and the mass fractionation factor per atomic mass unit of $0.10\%\pm0.03\%$.

The true $^{206}Pb/^{204}Pb$ (α) is related to the measured $^{206}Pb/^{204}Pb$ (α_m) and the linear mass fractionation factor (ε) (see Chapter 10 for mass fractionation laws) by

$$\alpha = \alpha_m(1+2\varepsilon).$$

Since $\alpha_m = 18.792$, $\sigma_{\alpha_m} = 0.005$, $\varepsilon = 0.001$, $d\varepsilon = 0.0003$, then

$$\alpha = \alpha_m(1+2\varepsilon) = 18.792*(1+2*0.001) = 18.830.$$

The partial derivative of α with respect to α_m is

$$\frac{\partial\alpha}{\partial\alpha_m} = 1+2\varepsilon.$$

The partial derivative of α with respect to ε is

$$\frac{\partial\alpha}{\partial\varepsilon} = 2\alpha_m.$$

If the uncertainties in α_m and ε are independent and random, then the uncertainty in α is

$$\sigma_a = \sqrt{\left(\frac{\partial\alpha}{\partial\alpha_m}\sigma_{a_m}\right)^2 + \left(\frac{\partial\alpha}{\partial\varepsilon}\sigma_\varepsilon\right)^2}$$

$$= \sqrt{\left[(1+2\varepsilon)\sigma_{a_m}\right]^2 + (2\alpha_m\sigma_\varepsilon)^2}$$

$$= \sqrt{\left[(1+2*0.001)*0.005\right]^2 + (2*18.792*0.0003)^2}$$

$$= 0.012.$$

Thus the true $^{206}Pb/^{204}Pb$ is $\alpha = 18.830\pm0.012$. Note that the $^{206}Pb/^{204}Pb$ error of 0.012 is significantly higher than the measured $^{206}Pb/^{204}Pb$ error of 0.005.

The maximum error can be calculated by

$$\sigma_\alpha = |\partial\alpha/\partial\alpha_m|\sigma_{\alpha_m} + |\partial\alpha/\partial\varepsilon|\sigma_\varepsilon = |1+2\varepsilon|\sigma_{\alpha_m} + |2\alpha_m|\sigma_\varepsilon$$

$$= |1+2*0.001|*0.005 + |2*18.792|*0.0003 = 0.016.$$

Example. Estimate the error of $^{207}Pb/^{206}Pb$ from measured $^{207}Pb/^{206}Pb$ and mass fractionation factor, provided that measured common Pb $^{207}Pb/^{206}Pb$ is $\omega_m = 0.8566 \pm 0.0002$ and the mass fractionation factor is $\varepsilon = 0.10\% \pm 0.03\%$.

The true $^{207}Pb/^{206}Pb$ (ω) is related to the measured $^{207}Pb/^{206}Pb$ (ω_m) and the mass fractionation factor (ε) by

$$\omega = (1+\varepsilon)\omega_m .$$

And the best estimate is

$$\omega = (1+\varepsilon)\omega_m = (1+0.001) \times 0.8566 = 0.8575 .$$

The partial derivative of ω relative to ω_m is

$$\frac{\partial \omega}{\partial \omega_m} = 1+\varepsilon .$$

And the partial derivative of ω relative to ε is

$$\frac{\partial \omega}{\partial \varepsilon} = \omega_m .$$

If the uncertainties in α_m and ε are independent and random, then the uncertainty in ω is

$$\sigma_\omega = \sqrt{\left(\frac{\partial \omega}{\partial \omega_m}\sigma_{\omega_m}\right)^2 + \left(\frac{\partial \omega}{\partial \varepsilon}\sigma_\varepsilon\right)^2} = \sqrt{\left[(1+\varepsilon)\sigma_{\omega_m}\right]^2 + \left(\omega_m \sigma_\varepsilon\right)^2}$$

$$= \sqrt{\left[(1+0.001)*0.0002\right]^2 + \left(0.8566*0.0003\right)^2}$$

$$= 0.0003 .$$

Therefore, the true $^{207}Pb/^{206}Pb$ is

$$\omega = 0.8575 \pm 0.0003 .$$

Example. Sometimes the uncertainty in the decay constant needs to be considered for the age calculations (Mattinson, 1987; Ludwig, 2000). Estimate the age error from $^{206}Pb/^{238}U$ in zircons provided $^{206}Pb/^{238}U = 0.5125 \pm 0.0002$ and the decay constant $\lambda_{238} = 1.551 \times 10^{-10}$ y^{-1} (year^{-1}) error of 0.1%.

The age equation is

$$t_{206} = \frac{1}{\lambda_{238}} \ln\left(\frac{^{206}Pb^*}{^{238}U} + 1\right) = \frac{1}{\lambda_{238}} \ln\left(R_{206} + 1\right),$$

where $R_{206} = {}^{206}Pb^*/{}^{238}U$.

The estimate of the zircon age is

$$t_{206} = \frac{1}{1.551 \times 10^{-10}} \ln\left(0.5125 + 1\right)$$
$$= 2.668 \times 10^9 \text{ y}.$$

The partial derivative of t_{206} with respect to R_{206} is

$$\frac{\partial t_{206}}{\partial R_{206}} = \frac{1}{\lambda_{206}(R_{206} + 1)}.$$

And the partial derivative of t_{206} with respect to λ_{238} is

$$\frac{\partial t_{206}}{\partial \lambda_{238}} = -\lambda_{238}^{-2} \ln\left(R_{206} + 1\right).$$

Since the uncertainties in R_{206} and λ_{238} are independent and random, then the uncertainty in the age of the zircon can be calculated using error propagation for independent random errors as

$$\sigma_{t_{206}} = \sqrt{\left[\frac{\sigma_{R_{206}}}{\lambda_{238}(R_{206} + 1)}\right]^2 + \left[-\sigma_{\lambda_{238}} \lambda_{238}^{-2} \ln(R_{206} + 1)\right]^2}$$

$$= \frac{1}{\lambda_{238}} \sqrt{\left(\frac{\sigma_{R_{206}}}{R_{206} + 1}\right)^2 + \left[\frac{\sigma_{\lambda_{238}}}{\lambda_{238}} \ln(R_{206} + 1)\right]^2}$$

$$= \frac{1}{1.551 \times 10^{-10}} \sqrt{\left(\frac{0.002}{0.5125 + 1}\right)^2 + \left[0.001 \times \ln(0.5125 + 1)\right]^2}$$

$$= 3 \times 10^6 \text{ y}.$$

Therefore, the zircon age is

$$t_{206} = 2.668 \pm 0.003 \times 10^9 \text{ y}.$$

References

Cheng, H., Edwards, R.L., Murrell, M.T. and Benjamin, T.M., 1998. Uranium-thorium-protactinium dating systematics. Geochim. Cosmochim. Acta, 62: 3437-3452.

Holmes, A., 1946. An estimate of the age of the Earth. Nature, 157: 680-684.

Houtermans, F.G., 1946. Die Isotopen-Haufigkeiten im naturlichen Blei und das Alter des Urans. Naturwissenschaften, 33: 185-187.

Ivanovich, M. and Harmon, R.S. (Editors), 1992. Uranium-series disequilibrium: Applications to Earth, marine, and environmental sciences. Oxford Science Publications, 910 pp.

Ku, T.L., 1968. Protactinium-231 methd of dating coral from Barbados Island. J. Geophys. Res., 73: 2271-2276.

Ludwig, K., 1998. On the treatment of concordant uranium-lead ages. Geochim. Cosmochim. Acta, 62: 665-676.

Ludwig, K.R., 1980. Calculation of uncertainties of U-Pb isotope data. Earth Planet. Sci. Lett., 46: 212-220.

Ludwig, K.R., 2000. Decay constant errors in U-Pb concordia-intercept ages. Chem. Geol., 166: 315-318.

Ludwig, K.R., 2003. User's Manual for ISOPLOT 3.00: A geochronological toolkit for Microsoft Excel. Berkeley Geochronology Center Special Publication No. 4, Berkeley, 70 pp.

Mattinson, J.M., 1987. U-Pb ages of zircons: A basic examination of error propagation. Chem. Geol., 66: 151-162.

Stacey, J.S. and Kramers, J.D., 1975. Approximation of terrestrial lead isotope evolution by a two-stage model. Earth Planet. Sci. Lett., 26: 207-221.

Tera, F. and Wasserburg, G.J., 1972. U-Th-Pb systematics in three Apollo 14 basalts and the problem of initial Pb in lunar rocks. Earth Planet. Sci. Lett., 14: 281-304.

Titterington, D.M. and Halliday, A.N., 1979. On the fitting of parallel isochrons and the method of maximum likelihood. Chem. Geol., 26: 183-195.

Wetherill, G.W., 1956. An interpretation of the Rhodesia and Witwaterand age patterns. Geochim. Cosmochim. Acta, 9: 290-292.

Chapter 13

Geochemical Kinematics and Dynamics

This chapter at first is concerned with the formulation of two fundamental dynamic and kinematic processes, diffusion and advection, for understanding elemental movement in various geochemical systems. Subsequently we will try to describe the bubble growth process using diffusion and advection equations that can help understand the magma eruption processes. We then talk about a kinematic problem: the trajectory of a volcanic bomb during volcanic eruption. Although this trajectory problem is not strictly geochemical, problems like this are indeed useful for understanding the physical aspects of the eruption process. A better understanding of physical process can improve our understanding of chemical process, which is the spirit and the essence of chemical geodynamics (Allegre, 1982; Zindler and Hart, 1986). In the end, we will discuss error function that is critical for the solutions of one-dimensional diffusion equations.

13.1. Diffusion

13.1.1. Formulation

The first Fick's law states that the flux (J) of diffusion is proportional to the concentration gradient (dc/dx)

$$J = -D\frac{dc}{dx},$$ (13.1)

where D is the diffusion coefficient of the element under consideration and is usually expressed in surface unit per time unit (e.g., cm^2/s), c the concentration of the element, and x is the coordinate. The negative sign indicates that the flux moves from high concentration to low concentration.

The second Fick's law of diffusion considers time dependent diffusion

$$\frac{dc}{dt} = -\frac{\partial J}{\partial x}. \tag{13.2}$$

Substitution of the first Fick's law of diffusion into Eq. (13.2) results in

$$\frac{dc}{dt} = \frac{\partial}{\partial x}\left(D\frac{\partial c}{\partial x}\right). \tag{13.3}$$

More details of the derivation of the diffusion/thermal equation can be found in Maaloe (1985) and Schubert et al. (2001).

If the diffusion coefficient D is independent of the composition and spatial position, then the diffusion equation becomes

$$\frac{dc}{dt} = D\left(\frac{\partial^2 c}{\partial x^2}\right). \tag{13.4}$$

Equation (13.4) describes the one-dimensional diffusion process. For a three-dimensional diffusion problem, the diffusion equation is

$$\frac{dc}{dt} = D\left(\frac{\partial^2 c}{dx^2} + \frac{\partial^2 c}{dy^2} + \frac{\partial^2 c}{dz^2}\right). \tag{13.5}$$

The solution to the one-dimensional diffusion Eq. (13.4) depends on boundary conditions.

1) For boundary conditions

$$c = c_1 \text{ at } t = 0 \text{ for } -h \leq x \leq h, \text{ and}$$

$$c = c_0 \text{ at } t = 0 \text{ for } |x| > h,$$

the solution of (13.4) that satisfies the above boundary conditions is

$$\frac{c - c_0}{c_1 - c_0} = \frac{1}{2}\left[erf\left(\frac{h-x}{2\sqrt{Dt}}\right) + erf\left(\frac{h+x}{2\sqrt{Dt}}\right)\right], \tag{13.6}$$

where *erf* is the error function. The details of the error function are provided at the end of this chapter (Appendix 13A). Note that if $c_0 = 0$, Eq. (13.6) is reduced to Eq. XI-34 in Maaloe (1985).

Example. Given $c_1 = 110$ ppm, $c_0 = 10$ ppm, if $D = 5 \times 10^{-11}$ cm^2/s and $h = 5$ cm, calculate the concentration at $x = 1$ cm at $t = 10^5$ years owing to diffusion.

The boundary condition in the example is similar to the intrusion of a small dike into a wall rock. The parameters can be shown in Fig. 13.1.

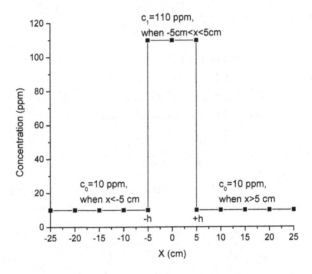

Fig. 13.1. The boundary conditions for the diffusion example before diffusion process starts (at *t*=0).

We at first try to calculate the two terms in the error function of Eq. (13.6):

$$\frac{h-x}{2\sqrt{Dt}} = \frac{5-1}{2\sqrt{5 \times 10^{-11} \times 10^5 \times 365 \times 24 \times 60 \times 60}} = 0.16,$$

$$\frac{h-x}{2\sqrt{Dt}} = \frac{5+1}{2\sqrt{5 \times 10^{-11} \times 10^5 \times 365 \times 24 \times 60 \times 60}} = 0.24.$$

According to the error function Table 13.1 in Appendix 13A, we have

$$Erf(0.16) = 0.179,$$

$$Erf(0.24) = 0.266.$$

Substitution of these values into Eq. (13.6), we obtain

$$\frac{c-10}{110-10} = \frac{1}{2}\left[erf(0.16) + erf(0.24) \right] = 0.222. \tag{13.7}$$

From Eq. (13.7), we find

$$c = 32.2 \text{ ppm},$$

at $x = 1$ cm at $t = 10^5$ years .

Thus the concentration at $x = 1$ cm decreases from 110 ppm to 32 ppm after 100,000 years due to diffusion.

2) For boundary conditions

$$c = c_1 \text{ at } t = 0 \text{ for } x > 0, \text{ and}$$

$$c = c_0 \text{ at } t = 0 \text{ for } x < 0,$$

the solution of Eq. (13.4) that satisfies the above boundary conditions is

$$\frac{c-c_0}{c_1-c_0} = \frac{1}{2} + \frac{1}{2} erf\left(\frac{x}{2\sqrt{Dt}} \right). \tag{13.8}$$

3) For boundary conditions

$$c = c_1 \text{ at } t = 0 \text{ for } x < 0, \text{ and}$$

$$c = c_0 \text{ at } t = 0 \text{ for } x > 0.$$

the solution of Eq. (13.4) that satisfies the above boundary conditions is

$$\frac{c-c_0}{c_1-c_0} = \frac{1}{2} - \frac{1}{2} erf\left(\frac{x}{2\sqrt{Dt}} \right). \tag{13.9}$$

4) For boundary conditions

$$c = c_1 \text{ at all } t \text{ for } x < 0, \text{ and}$$

$$c = c_0 \text{ at } t = 0 \text{ for } x > 0.$$

The solution of Eq. (13.4) that satisfies the above boundary conditions for $x > 0$ and $t > 0$ is

$$c = c_1 + (c_0 - c_1)erf\left(\frac{x}{2\sqrt{Dt}}\right). \qquad (13.10)$$

5) For radial diffusion, the diffusion Eq. (13.4) may be re-expressed as

$$\frac{dc}{dt} = D\left(\frac{\partial^2 c}{\partial r^2} + \frac{2}{r}\frac{\partial c}{\partial r}\right) = \frac{1}{r^2}\frac{\partial}{\partial r}\left(Dr^2\frac{\partial C}{\partial r}\right), \qquad (13.11)$$

with the following boundary conditions:

$$c = c_1 \text{ at } t = 0 \text{ for } r < a, \text{ and}$$

$$c = c_0 \text{ at all } t \text{ for } r > a,$$

where a is the radius of the sphere.

The solution of Eq. (13.11) that satisfies the above boundary conditions is

$$\frac{c - c_0}{c_1 - c_0} = 1 + \frac{2a}{\pi r}\sum_1^\infty \frac{(-1)^n}{n}\sin\left(\frac{n\pi r}{a}\right)\exp\left(\frac{-Dn^2\pi^2 t}{a^2}\right). \qquad (13.12)$$

More detailed mathematical treatments of the diffusion process can be found in Crank (1975).

13.1.2. Variation of diffusion coefficient

The diffusion coefficient varies with temperature and pressure. The temperature dependence is expressed by the following expression first determined by Arrhenius

$$D = D_0 \exp\left(-\frac{E}{RT}\right), \qquad (13.13)$$

where E is the activation energy, R is the gas constant, T the absolute temperature and D_0 is frequency factor.

By taking the logarithm of both sides of Eq. (13.13), we get

$$\log D = \log D_0 - \frac{E}{2.303 RT}. \qquad (13.14)$$

Both the activation energy (E) and the frequency factor (D_0) can be extracted from a linear array of diffusion data by plotting experimentally-determined $\log D$ as a function of $1/T$.

The pressure dependence of the diffusion coefficient is given by

$$D = D_0 \exp\left(-\frac{PV}{RT}\right), \qquad (13.15)$$

where V is the activation volume, R is the gas constant, T the absolute temperature and D_0 is frequency factor.

When both temperature and pressure are considered, the modified Arrhenius equation is (Harrison et al., 1985)

$$D = D_0 \exp\left(-\frac{E + PV}{RT}\right). \qquad (13.16)$$

13.2. Advection and Percolation

In a frame fixed to the matrix, the mass conservation equation for the solid and melt in a one-dimensional melting column is (McKenzie, 1984; Richter and McKenzie, 1984; Navon and Stolper, 1987)

$$\rho_f \phi \frac{\partial C_f}{\partial t} + (1 - \phi)\rho_s \frac{\partial C_s}{\partial t} + v_f \frac{\partial C_f}{\partial z} = 0, \qquad (13.17)$$

where C_f is the concentration of an element in the fluid, C_s is the concentration of the element in the solid, t is time, z is the distance traversed along the column, ρ_f is the density of the fluid, ρ_s is the density of the solid, v_f is fluid velocity, and ϕ is the fluid volume fraction in the column. Note that the diffusion term is ignored in Eq. (13.17).

Assuming complete local chemical equilibrium between the solid and melt ($C_s = K_d C_f$), Eq. (13.17) may be written as

$$\frac{\partial C_f}{\partial t} + v_i \frac{\partial C_f}{\partial z} = 0, \qquad (13.18)$$

where

$$v_i = \frac{\rho_f \phi}{\rho_f \phi + \rho_f (1-\phi) K_d} v_f .$$

(13.19)

v_i is the rate at which a point of constant concentration for an element moves through the column, that is

$$v_i = \left(\frac{\partial z}{\partial t} \right)_{C_f} .$$

(13.20).

Since $\rho_f (1-\phi) K_d > 0$, we have $v_i < v_f$. That is, trace element concentrations move slower than the fluid itself. Since compatible elements have higher K_d values than incompatible elements, on the basis of Eq. (13.19), compatible elements are transported at lower velocities than incompatible ones.

13.3. Bubble Growth by Mass Transfer of Volatiles

13.3.1. Formulation

Growth of bubble by the mass transfer of volatiles from the magma to the bubble in a sphere coordinate is given by (Proussevitch and Sahagian, 1998)

$$\frac{\partial C}{\partial t} + v \frac{\partial C}{\partial r} = \frac{1}{r^2} \frac{\partial}{\partial r} \left(D r^2 \frac{\partial C}{\partial r} \right).$$

(13.21)

Note that the second term on the left hand side is the advection term (see Eq. (13.18)) and the term on the right hand side is the diffusion term (see Eq. (13.11)).
The overpressure of the bubble is given by (Proussevitch et al., 1993)

$$P_g - P_e = \frac{2\sigma}{R} + 4\eta v_R \left(\frac{1}{R} - \frac{R^2}{S^3} \right).$$

(13.22)

where p_g is the internal pressure of the bubble, p_e is the external pressure, σ is the surface tension, R is the radius of the bubble, S is the radius of the melt shell enclosing the bubble.
The boundary conditions are:

at the interface of bubble-melt internal shell, we have

$$\frac{d}{dt}\left(\frac{4}{3}\pi\rho_g R^3\right) = 4\pi R^2 \rho_m D\left(\frac{\partial C}{\partial r}\right)_R ; \qquad (13.23)$$

and at the outside boundary of the melt shell, we have

$$\left(\frac{\partial C}{\partial r}\right)_S = 0 . \qquad (13.24)$$

The relationship between v in Eq. (13.21) and the bubble growth rate v_R in Eq. (13.22) can be obtained on the basis of continuity equation for spherical coordinates:

$$\frac{1}{r}\frac{\partial}{\partial r}(r^2 v) = 0 , \qquad (13.25)$$

thus $r^2 v$ is independent of r but is a function of t :

$$r^2 v = f(t) , \qquad (13.26)$$

or

$$v = f(t)/r^2 . \qquad (13.27)$$

The particle velocity at the bubble interface $r = R(t)$ is v_R, which is he bubble growth rate. Using this boundary condition, from Eq. (13.27), we have

$$v_R = f(t)/R^2 . \qquad (13.28)$$

Combination of Eqs. (13.27) and (13.28) yields

$$v = \frac{R^2}{r^2} v_R . \qquad (13.29)$$

Since v_R is the growth rate of the bubble ($v_R = R' = dR/dt$), Eq. (13.29) can be written as

$$v = \frac{R^2}{r^2} v_R = \frac{R^2}{r^2}\frac{dR}{dt} \qquad (13.30)$$

The combination of Eq. (13.21) and Eq. (13.30) gives

$$\frac{\partial C}{\partial t} + v_R \frac{r^2}{R^2} \frac{\partial C}{\partial r} = \frac{1}{r^2} \frac{\partial}{\partial r} \left(Dr^2 \frac{\partial C}{\partial r} \right),$$ (13.31)

where $R < r < S$.

The system of Eqs. (13.22) and (13.31) with boundary conditions (13.23) and (13.24) can be solved numerically.

13.3.2. Quasi-static approximation

When the Pecet number, the measure of the relative importance of advection to diffusion, is small, which is the case for high viscosity magmas, the temporal derivation of concentration and the advection term in Eq. (13.31) may be ignored, and quasi-static approximation may be developed. In this case, Eq. (13.31) reduces to

$$\frac{1}{r^2} \frac{\partial}{\partial r} \left(Dr^2 \frac{\partial C}{\partial r} \right) = 0 .$$ (13.32)

The solution of (13.32) is

$$C = A - \frac{B}{r},$$ (13.33)

where A and B are constants that can be determined from boundary conditions.

From Eq. (13.33) the concentration at the bubble internal surface (at $r = R$) is

$$C_R = A - \frac{B}{R},$$ (13.34)

or

$$A = C_R + \frac{B}{R} .$$ (13.35)

Conservation of water mass in the spherical shell of melt that surrounds the bubble at $r = S$ (the second boundary condition) requires

$$\frac{4}{3} \pi \rho_g R^3 + \int_R^S 4\pi r^2 \rho_m C dr = \frac{4}{3} \pi \rho_m S_0^3 C_0 .$$ (13.36)

Substitution of Eq. (13.33) into Eq. (13.36) gives

$$\frac{4}{3}\pi\rho_g R^3 + \frac{4}{3}\pi\rho_m \left(S^3 - R^3\right)\left(C_R + \frac{B}{R}\right) - 2\pi\rho_m \left(S^2 - R^2\right)B$$
$$= \frac{4}{3}\pi\rho_m S_0^{\ 3} C_0.$$

(13.37)

From Eq. (13.37) we obtain

$$B = \frac{S_0^{\ 3}\left(C_0 - C_R\right) - \dfrac{\rho_g}{\rho_m}R^3}{\dfrac{S_0^{\ 3}}{R} - 1.5\left[\left(S_0^{\ 3} + R^3\right)^{2/3} - R^2\right]} .$$

(13.38)

Substitution of Eq. (13.38) into Eq. (13.33) and differentiation of Eq. (13.33) with respect to r gives

$$\left(\frac{\partial C}{\partial r}\right)_R = \frac{B}{R^2} = \frac{S_0^{\ 3}\left(C_0 - C_R\right) - \dfrac{\rho_g}{\rho_m}R^3}{S_0^{\ 3}R - 1.5\left[\left(S_0^{\ 3} + R^3\right)^{2/3} - R^2\right]R^2} .$$

(13.39)

dR/dt can be obtained by rewriting Eq. (13.22) as

$$\frac{dR}{dt} = \frac{1}{4\eta}\left(\frac{GT}{M}\rho_g - P_f - \frac{2\sigma}{R}\right)\frac{S_0^{\ 3} + R^3}{S_0^{\ 3}}R,$$

(13.40)

where the term $\dfrac{GT}{M}\rho_g$ is the gas pressure in the bubble (p_g). ρ_g is the density of gas, M is the molecular weight of water, G is the universal gas constant, and T is temperature.

When $S \gg R$ (thick melt shell), and ignoring the surface tension σ, then Eq. (13.40) reduces to

$$\frac{dR}{dt} = \frac{1}{4\eta}\left(\frac{GT}{M}\rho_g - P_f\right)R.$$

(13.41)

Similarly, Eq. (13.39) reduces to

$$\left(\frac{\partial C}{\partial r}\right)_R = \frac{(C_0 - C_R)}{R}. \tag{13.42}$$

We also need to find the variation of ρ_g in the bubble. $d\rho_g/dt$ can be obtained by rewriting Eq. (13.23) as:

$$\frac{d\rho_g}{dt} = \frac{3D\rho_m}{R}\left(\frac{\partial C}{\partial r}\right)_R - \frac{3\rho V_R}{R}. \tag{13.43}$$

Since $v_R = dR/dt$, substitution of Eq. (13.42) into Eq. (13.43) results in

$$\frac{d\rho_g}{dt} = \frac{3D\rho_m(C_0 - C_R)}{R^2} - \frac{3\rho}{R}\frac{dR}{dt}. \tag{13.44}$$

The asymptotic solution to the system of Eqs. (13.41) and (13.44) is given by (Lyakhovsky et al., 1996)

$$R^2 = \frac{2D\rho_m(C_0 - C_R)}{\rho_g}t - \frac{2}{3}\frac{D\eta}{P_f}\frac{\rho_m}{\rho_g}(2C_0 + C_R)\log\left(\frac{\Delta P}{\eta}t\right). \tag{13.45}$$

13.4. Bubble Growth by Decompression

13.4.1. Thin shell of over-pressured viscous melt

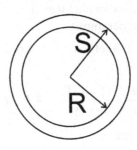

Fig. 13.1. A diagram for a thin shell of melt enclosing a bubble with radius R. The thickness of the thin shell is $(S - R)$ and is significantly less than R.

The growth of a bubble in a thin shell of over-pressured viscous melt (Fig. 13.1) is given by (Barclay et al., 1995)

$$p_g - p_e = \frac{12\eta hR'}{R^2}, \tag{13.46}$$

where p_g is the internal gas pressure of the bubble, p_e the external pressure, R is the radius of the bubble, η is the dynamic viscosity, h the thickness of the bubble and $h = (S - R)$, R' is the growth rate of the bubble and is thus dR/dt.

Equation (13.46) can be rewritten as

$$dR/dt = R^2 (p_g - p_e)/(12\eta h). \tag{13.47}$$

To solve Eq. (13.47), we also need to express p_g and h as a function of R or t. Assuming perfect gas law and isothermal conditions, the internal gas pressure (p_g) and the radius of the bubble (R) are related to the initial gas pressure and initial bubble radius before decompression by (Sparks, 1978)

$$P_0 R_0^{\,3} = P_g R^3, \tag{13.48}$$

or

$$p_g = P_0 R_0^{\,3} / R^3. \tag{13.49}$$

The volume of the shell is given by

$$V = \frac{4}{3}\pi (R+h)^3 - \frac{4}{3}\pi R^3$$
$$= \frac{4}{3}\pi \left(3R^2 h + 3Rh^2 + h^3\right). \tag{13.50}$$

For thin shell, $h \ll R$, Eq. (13.50) may be approximated as

$$V \approx \frac{4}{3}\pi h^3. \tag{13.51}$$

Assuming constant density of the melt shell, the constant mass of the melt shell gives the following relationship:

$$4\pi R^2 h = 4\pi R_0^{\,2} h_0, \tag{13.52}$$

or

$$h = R_0^2 h_0 / R^2 .$$ (13.53)

Substitution of Eqs. (13.49) and (13.53) into (13.47) yields

$$\frac{dR}{dt} = \frac{R^4 \left(\frac{p_0 R_0^3}{R^3} - P_e \right)}{12 \eta R_0^2 h_0} .$$ (13.54)

As far as the external pressure (p_e) is concerned, thee are two simple cases: (1) instantaneous decompression when p_e is a constant, and (2) linear decompression with constant rate of decompression. The first case might occur in a sudden volcanic explosion or when a magma column is decompressed by sector collapse. The second case is an approximation to pressure evolution during magma ascent.

13.4.1.1. Instantaneous decompression

When p_e is a constant, the solution to Eq. (13.54) is (Barclay et al., 1995)

$$t = \frac{4 \eta h_o}{p_o R_o} \ln \left[\frac{R^3}{R_o^{\,3}} \left(1 - \frac{p_e}{p_o} \right) \bigg/ \left(1 - \frac{p_e R^3}{p_o R_o^{\,3}} \right) \right],$$ (13.55)

or

$$R = R_o \left[\frac{p_e}{p_o} + \left(1 - \frac{p_e}{p_o} \right) \exp \left(\frac{-p_0 R_0 t}{4 \eta h_o} \right) \right]^{-1/3} .$$ (13.56)

13.4.1.2. Linear decompression

In many circumstances during magma ascent the pressure decreases at a finite rate. Assuming the rate of decompression is a constant, the consequences for linear decompression using the thin shell model may be presented to express the bubble radius as a function of time in a constantly varying pressure field. The external pressure is expressed by

$$p_e = \rho g (d_o - wt) .$$ (13.57)

where ρ is the density of the magma body, g the acceleration due to gravity, w velocity of ascent, and d_o initial depth of the ascending bubble. Substitution of Eq. (13.57) into (13.54) gives

$$\frac{dR}{dt} = \frac{R^4\left[\frac{P_0 R_0^3}{R^3} - \rho g(d_0 - wt)\right]}{12\eta R_0^2 h_0}.$$ (13.58)

Let

$$s = (R_o / R)^3.$$

Equation (13.58) may be written as

$$\frac{ds}{dt} + \frac{R_o p_o}{4\eta h_o} s = \frac{R_o p_o}{4\mu h_o} \frac{\rho g d_o - \rho g w t}{p_o}.$$ (13.59)

The solution to Eq. (13.59) with initial condition $R(0) = R_0$ is (Barclay et al., 1995)

$$R = R_o\left[\left(1 - b - \frac{c}{a}\right)e^{at} + b + \frac{c}{a} - ct\right]^{-1/3},$$ (13.60)

where

$$a = P_o R_o /(4\eta h_o),$$

$$b = \rho g d_o / p_o,$$

$$c = \rho g w / p_o.$$

13.4.2. Thick shell melt model during instantaneous decompression

For a thick shell model (Fig. 13.2) where $R \ll S$, we have $R^2 / S^3 \ll 1/R$. If we further ignore the surface tension (σ), then Eq. (13.22) may be simplified as

$$P_g - P_e = \frac{4\eta v_R}{R}.$$ (13.61)

Note that the bubble growth rate v_R is

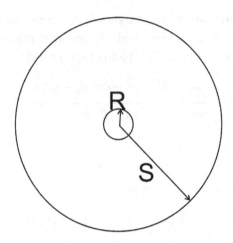

Fig. 13.2. A thick shell melt model where the thickness of the melt shell $(S - R)$ is significantly greater than the radius of the gas bubble R.

$$v_R = \frac{dR}{dt}.$$

Equation (13.61) may be written as

$$P_g - P_e = \frac{4\eta}{R}\frac{dR}{dt}. \qquad (13.62)$$

By assuming perfect gas law and by substituting (13.49) into Eq. (13.62), we get

$$\frac{P_0 R_0^{\,3}}{R^3} - P_e = \frac{4\eta}{R}\frac{dR}{dt}. \qquad (13.63)$$

During instantaneous decompression, the external pressure (P_e) is a constant. The solution of Eq. (13.63) with initial condition $R(0) = R_0$ is given by (Barclay et al., 1995)

$$R = R_0 \left\{ \frac{P_0}{P_e}\left[1 - \exp\left(\frac{-3P_e t}{4\eta}\right)\right] + \exp\left(\frac{-3P_e t}{4\eta}\right) \right\}^{\frac{1}{3}}. \qquad (13.64)$$

13.5. The Trajectories of a Volcanic Bomb

13.5.1. No drag

The initial horizontal velocity is u_0 and the initial vertical velocity is w_0. Thus the angel of the initial trajectory is

$$\tan \beta = \frac{w_0}{u_0}. \tag{13.65}$$

When there is no drag, for horizontal velocity, we have

$$\frac{du}{dt} = 0, \tag{13.66}$$

and for vertical velocity, we have

$$\frac{dw}{dt} = -g, \tag{13.67}$$

where g is the standard acceleration of free fall.

The solutions of Eqs. (13.66) and (13.67) are

$$u = u_0, \tag{13.68}$$

$$w = -gt + w_0. \tag{13.69}$$

Let x be the distance from the initial eruption and z be the vertical distance from the initial eruption, on the basis of Eqs. (13.68) and (13.69) , we have

$$\frac{dx}{dt} = u_0, \tag{13.70}$$

$$\frac{dz}{dt} = -gt + w_0. \tag{13.71}$$

The solutions to Eqs. (13.70) and (13.71) are

$$x = u_0 t, \tag{13.72}$$

$$z = -\frac{1}{2}gt^2 + w_0 t. \tag{13.73}$$

The combination of Eqs. (13.72) and (13.73) gives

$$z = -\frac{1}{2}\frac{g}{u_0^2}x^2 + \frac{w_0}{u_0}x \,. \tag{13.74}$$

Problem. How far can a volcanic bomb go?

Assuming that the volcanic bomb falls at $z = 0$ (the final vertical elevation is identical to the initial elevation), then Eq. (13.74) is reduced to

$$\frac{1}{2}\frac{g}{u_0^2}x^2 - \frac{w_0}{u_0}x = 0 \,. \tag{13.75}$$

The nontrivial solution to Eq. (13.75) is

$$x = \frac{2u_0 w_0}{g} \,. \tag{13.76}$$

If $z < 0$ (for example, the volcanic bomb falls on a ground that is lower than the initial eruption elevation), then we obtain two solutions from the quadratic equation (13.74)

$$x = \frac{w_0 \pm \sqrt{w_0 - 2gz}}{g/u_0} \,. \tag{13.77}$$

For $z < 0$, we have $w_0 < \sqrt{w_0 - 2gz}$. In order for $x > 0$, only one solution from Eq. (13.77) is practical:

$$x = \frac{w_0 + \sqrt{w_0 - 2gz}}{g/u_0} \,. \tag{13.78}$$

Example. If initial horizontal velocity is 30 m/s, and initial vertical velocity is 30 m/s, calculate the distance that the volcanic bomb can go. The standard acceleration of free fall is 9.8 m/s^2.

Since $u_0 = 30$ m/s and $w_0 = 30$ m/s, according to Eq. (13.76), the horizontal distance is

$$x = \frac{2 \times 30 \times 30}{9.8} = 183.7 \text{ m} \,.$$

The initial eruption angel is

$$\tan \beta = \frac{30}{30} = 1 \, ,$$

and thus $\beta = 45°$.

13.5.2. Linear drag

Assuming that the resistance is linearly proportional to the velocity, then for the variation of the horizontal velocity, we have (Middleton and Wilcock, 1994)

$$\frac{du}{dt} = -k_1 u \, . \tag{13.79}$$

Equation (13.79) may be written

$$\frac{du}{u} = -k_1 dt \, . \tag{13.80}$$

The solution to Eq. (13.80) is

$$\ln u = -k_1 t + C_1 \, . \tag{13.81}$$

From initial condition,

$$u = u_0 \text{ at } t = 0 \, ,$$

we obtain

$$C_1 = \ln u_0 \, . \tag{13.82}$$

The combination of Eq. (13.81) with Eq. (13.82) yields

$$u = u_0 e^{-k_1 t} \, . \tag{13.83}$$

For the variation of the vertical velocity, we have

$$\frac{dw}{dt} = -g - k_2 w \, . \tag{13.84}$$

The solution to Eq. (13.84) is

$$w = -\frac{g}{k_2} + C_2 e^{-k_2 t} \, . \tag{13.85}$$

From initial condition,

$$w = w_0 \text{ at } t = 0 \, ,$$

we get

$$C_2 = w_0 + \frac{g}{k_2}. \tag{13.86}$$

Substitution of Eq. (13.86) into (13.85) results in (Middleton and Wilcock, 1994)

$$w = -\frac{g}{k_2} + \left(w_0 + \frac{g}{k_2} \right) e^{-k_2 t}. \tag{13.87}$$

The horizontal distance is related to the horizontal velocity by

$$\frac{dx}{dt} = u = u_0 e^{-k_1 t}. \tag{13.88}$$

Integration of Eq. (13.88) with initial condition $x = 0$ at $t = 0$ gives

$$x = \frac{u_0}{k_1} \left(1 - e^{-k_1 t} \right). \tag{13.89}$$

The vertical distance is related to the vertical velocity by

$$\frac{dz}{dt} = w = -\frac{g}{k_2} + \left(w_0 + \frac{g}{k_2} \right) e^{-k_2 t}. \tag{13.90}$$

Integration of Eq. (13.90) with initial condition $z = 0$ at $t = 0$ yields

$$z = -\frac{g}{k_2} t + \frac{w_0 + \dfrac{g}{k_2}}{k_2} \left(1 - e^{-k_2 t} \right). \tag{13.91}$$

Equations (13.89) and (13.91) describe the projectile trajectories of a volcanic bomb for linear drag.

Appendix 13A. Error Function

Error function is useful in applied mathematics and engineering. The values of the error function,

$$erf(z) = \frac{2}{\sqrt{\pi}} \int_0^z e^{-u^2} du,$$

for different z are given in Table 13.1 and are plotted in Fig. 13.3.

Table 13.1. The error function $erf(z) = \dfrac{2}{\sqrt{\pi}} \displaystyle\int_0^z e^{-u^2} du$.

z	erf(z)	Z	erf(z)	z	erf(z)	Z	erf(z)	z	erf(z)
0.00	0.00000	0.50	0.52050	1.00	0.84270	1.50	0.96611	2.00	0.99532
0.01	0.01128	0.51	0.52924	1.01	0.84681	1.51	0.96728	2.01	0.99552
0.02	0.02256	0.52	0.53790	1.02	0.85084	1.52	0.96841	2.02	0.99572
0.03	0.03384	0.53	0.54646	1.03	0.85478	1.53	0.96952	2.03	0.99591
0.04	0.04511	0.54	0.55494	1.04	0.85865	1.54	0.97059	2.04	0.99609
0.05	0.05637	0.55	0.56332	1.05	0.86244	1.55	0.97162	2.05	0.99626
0.06	0.06762	0.56	0.57162	1.06	0.86614	1.56	0.97263	2.06	0.99642
0.07	0.07886	0.57	0.57982	1.07	0.86977	1.57	0.97360	2.07	0.99658
0.08	0.09008	0.58	0.58792	1.08	0.87333	1.58	0.97455	2.08	0.99673
0.09	0.10128	0.59	0.59594	1.09	0.87680	1.59	0.97546	2.09	0.99688
0.10	0.11246	0.60	0.60386	1.10	0.88021	1.60	0.97635	2.10	0.99702
0.11	0.12362	0.61	0.61168	1.11	0.88353	1.61	0.97721	2.11	0.99715
0.12	0.13476	0.62	0.61941	1.12	0.88679	1.62	0.97804	2.12	0.99728
0.13	0.14587	0.63	0.62705	1.13	0.88997	1.63	0.97884	2.13	0.99741
0.14	0.15695	0.64	0.63459	1.14	0.89308	1.64	0.97962	2.14	0.99753
0.15	0.16800	0.65	0.64203	1.15	0.89612	1.65	0.98038	2.15	0.99764
0.16	0.17901	0.66	0.64938	1.16	0.89910	1.66	0.98110	2.16	0.99775
0.17	0.18999	0.67	0.65663	1.17	0.90200	1.67	0.98181	2.17	0.99785
0.18	0.20094	0.68	0.66378	1.18	0.90484	1.68	0.98249	2.18	0.99795
0.19	0.21184	0.69	0.67084	1.19	0.90761	1.69	0.98315	2.19	0.99805
0.20	0.22270	0.70	0.67780	1.20	0.91031	1.70	0.98379	2.20	0.99814
0.21	0.23352	0.71	0.68467	1.21	0.91296	1.71	0.98441	2.21	0.99822
0.22	0.24430	0.72	0.69143	1.22	0.91553	1.72	0.98500	2.22	0.99831
0.23	0.25502	0.73	0.69810	1.23	0.91805	1.73	0.98558	2.23	0.99839
0.24	0.26570	0.74	0.70468	1.24	0.92051	1.74	0.98613	2.24	0.99846
0.25	0.27633	0.75	0.71116	1.25	0.92290	1.75	0.98667	2.25	0.99854
0.26	0.28690	0.76	0.71754	1.26	0.92524	1.76	0.98719	2.26	0.99861
0.27	0.29742	0.77	0.72382	1.27	0.92751	1.77	0.98769	2.27	0.99867
0.28	0.30788	0.78	0.73001	1.28	0.92973	1.78	0.98817	2.28	0.99874
0.29	0.31828	0.79	0.73610	1.29	0.93190	1.79	0.98864	2.29	0.99880
0.30	0.32863	0.80	0.74210	1.30	0.93401	1.80	0.98909	2.30	0.99886
0.31	0.33891	0.81	0.74800	1.31	0.93606	1.81	0.98952	2.31	0.99891
0.32	0.34913	0.82	0.75381	1.32	0.93807	1.82	0.98994	2.32	0.99897
0.33	0.35928	0.83	0.75952	1.33	0.94002	1.83	0.99035	2.33	0.99902
0.34	0.36936	0.84	0.76514	1.34	0.94191	1.84	0.99074	2.34	0.99906
0.35	0.37938	0.85	0.77067	1.35	0.94376	1.85	0.99111	2.35	0.99911
0.36	0.38933	0.86	0.77610	1.36	0.94556	1.86	0.99147	2.36	0.99915
0.37	0.39921	0.87	0.78144	1.37	0.94731	1.87	0.99182	2.37	0.99920
0.38	0.40901	0.88	0.78669	1.38	0.94902	1.88	0.99216	2.38	0.99924
0.39	0.41874	0.89	0.79184	1.39	0.95067	1.89	0.99248	2.39	0.99928
0.40	0.42839	0.90	0.79691	1.40	0.95229	1.90	0.99279	2.40	0.99931

0.41	0.43797	0.91	0.80188	1.41	0.95385	1.91	0.99309	2.41	0.99935
0.42	0.44747	0.92	0.80677	1.42	0.95538	1.92	0.99338	2.42	0.99938
0.43	0.45689	0.93	0.81156	1.43	0.95686	1.93	0.99366	2.43	0.99941
0.44	0.46623	0.94	0.81627	1.44	0.95830	1.94	0.99392	2.44	0.99944
0.45	0.47548	0.95	0.82089	1.45	0.95970	1.95	0.99418	2.45	0.99947
0.46	0.48466	0.96	0.82542	1.46	0.96105	1.96	0.99443	2.46	0.99950
0.47	0.49375	0.97	0.82987	1.47	0.96237	1.97	0.99466	2.47	0.99952
0.48	0.50275	0.98	0.83423	1.48	0.96365	1.98	0.99489	2.48	0.99955
0.49	0.51167	0.99	0.83851	1.49	0.96490	1.99	0.99511	2.49	0.99957
								2.50	0.99959
								2.60	0.99976
								2.70	0.99987
								2.80	0.99992
								2.90	0.99996
								3.00	0.99998

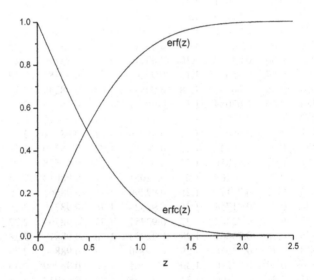

Fig. 13.3. Plot of error function. Note that $erfc(z) = 1 - erf(z)$.

Note that $erf(-z) = -erf(z)$. For computer calculations, the error function can be approximated by $erf(z) = 1 - \left(a_1 t + a_2 t^2 + a_3 t^3 \right) \exp\left[-\left(z^2 \right) \right]$ where $t = 1/(1 + 0.47047z)$, $a_1 = 0.3480242$, $a_2 = -0.0958798$, $a_3 = 0.7478556$. The error in this approximation is $< 2.5 \times 10^{-5}$.

References

Allegre, C.J., 1982. Chemical Geodynamics. Tectonophysics, 81: 109-132.

Barclay, J., Riley, D.S. and Sparks, R.S.J., 1995. Analytical models for bubble growth during decompression of high viscosity magmas. Bull. Volcanol., 57: 422-431.

Crank, J., 1975. The mathematics of diffusion. Claredon Press. Oxford, 414 pp.

Harrison, T.M., Duncan, I. and McDougall, I., 1985. Diffusion of ^{40}Ar in biotite: temperature, pressure and compositional effects. Geochim. Cosmochim. Acta, 49: 2461-2468.

Lyakhovsky, V., Hurwitz, S. and Navon, O., 1996. Bubble growth in rhyolitic melts: experimental and numerical investigation. Bull. Volcanol., 58: 19-32.

Maaloe, S., 1985. Principles of Igneous Petrology. Springer-Verlag, Berlin, Heidelberg, New York, Tokyo, 374 pp.

McKenzie, D.P., 1984. The generation and compaction of partial melts. J. Petrol., 25: 713-765.

Middleton, G.V. and Wilcock, P.R., 1994. Mechanics in the Earth and Environmental Sciences. Cambridge University Press, Cambridge, 459 pp.

Navon, O. and Stolper, E., 1987. Geochemical consequences of melt percolation: The upper mantle as a chromatographic column. J. Geology, 95: 285-307.

Proussevitch, A.A. and Sahagian, D.L., 1998. Dynamics and energetics of bubble growth in magmas: Analytical formulation and numerical modeling. J. Geophys. Res., 103: 18223-18251.

Proussevitch, A.A., Sahagian, D.L. and Anderson, A.T., 1993. Dynamics of diffusive bubble growth in magmas: isothermal case. J. Gephys. Res., 98: 283-307.

Richter, F.M., McKenzie, D., 1984. Dynamic models for melt segregation from a deformable matrix. J. Geology, 92: 729-740.

Schubert, G., Turcotte, D.L. and Olsen, P., 2001. Mantle convection in the Earth and Planet. Cambridge University Press, Cambridge, 940 pp.

Sparks, R.S.J., 1978. The dynamics of bubble formation and growth in magmas: A review and analysis. J. Volcanol. Geotherm. Res., 3: 1-37.

Zindler, A. and Hart, S.R., 1986. Chemical geodynamics. Ann. Rev. Earth Planet. Sci., 14: 493-571.